Edexcel

Functional Skills

Mathematics

Student Book
Level 2

Series Editor: Tony Cushen

Authors:
Chris Baston
Tony Cushen
Alistair Macpherson
Su Nicholson
Carol Roberts

Published by Pearson Education Limited, a company incorporated in England and Wales, having its registered office at Edinburgh Gate, Harlow, Essex, CM20 2JE. Registered company number: 872828

Edexcel is a registered trademark of Edexcel Limited

© Pearson Education Limited 2010

The rights of Chris Baston, Tony Cushen, Alistair Macpherson, Su Nicholson and Carol Roberts to be identified as the authors of this Work have been asserted by them in accordance with the Copyright, Designs and Patent Act, 1988.

First published 2010

13
10 9 8 7 6

British Library Cataloguing in Publication Data
A catalogue record for this book is available from the British Library

ISBN 978 1 84690 770 8

Typeset by Techset, Gateshead
Picture research by Katharine Oakes and Alison Prior
Printed in Malaysia (CTP-VP)

Acknowledgements

Picture Credits

The publisher would like to thank the following for their kind permission to reproduce their photographs:

(Key: b-bottom; c-centre; l-left; r-right; t-top)

Alamy Images: ColsMountains 37, G&D Images 20, Photoshot Holdings Ltd 12, MBI 28, PCL 29, Network Photographers 16, Angus Taylor 46, Art Directors & TRIP 93; **Corbis:** Creasource 99; **Getty Images:** Rita Maas / FoodPix 34tr; **iStockphoto:** 23tr, 98, 101, 107, Claude Beaubien 38, Doug Cannell 22, Robert Churchill 103, Joe Clemson 34cl, David H. Lewis 33, Felix Möckel 108, Frank Ramspott 106, Jeremy Richards 21br, SangHyunPaek 24tr, Edward Shaw 53; **Pearson Education Ltd:** MindStudio 14tr; **Shutterstock:** Oleinikova Olga 44-45, Andrejs Pidjass 44-45/2; **Thinkstock:** Comstock 63, iStockphoto 14br, 24br, 110, iStockphoto 14br, 24br, 110, Photodisc 31, Pixland 23br, Stockbyte 21tr

Cover images: *Front:* **Shutterstock:** Norma Cornes

All other images © Pearson Education

Every effort has been made to trace the copyright holders and we apologise in advance for any unintentional omissions. We would be pleased to insert the appropriate acknowledgement in any subsequent edition of this publication.

We are grateful to the following for permission to reproduce copyright material:

Figures
Figure on page 59 adapted from http://www.floorplanskitchen.com/floor-plans/kitchen-plans-with-peninsulas/ and http://www.ikea.com/ms/en_GB/rooms_ideas/splashplanners.html, © interIKEAsystem B.V. 2010; Figure on page 91 from Survey of family spending charts half century of consumer culture by Martin Hickman, http://www.independent.co.uk/news/uk/home-news/survey-of-family-spending-charts-half-century-of-consumer-culture-775185.html, © The Independent 2008

Tables
Table on page 13 adapted from http://en.wedur.is/climatology/clim/nr/1213, © Icelandic Meteorological Office; Table on page 49 adapted from http://www.eastcoast.co.uk, image courtesy of East Coast Trains; Table on page 88 adapted from http://www.h4u.co.uk/; Table on page 95 adapted from
http://www.premierleague.com/page/statistics/0,12306,00.html, with permission of Premier League © 2010

Screenshot
Screenshot on page 32 from Explorer 191 Banbury, Bicester & Chipping Norton (Brackley) 1:25 000 scale (ISBN 0-319-21822-8), Reproduced by permission of Ordnance Survey on behalf of HMSO (C) Crown copyright (2010). All rights reserved.
In some instances we have been unable to trace the owners of copyright material, and we would appreciate any information that would enable us to do so.

Every effort has been made to trace the copyright holders and we apologise in advance for any unintentional omissions. We would be pleased to insert the appropriate acknowledgement in any subsequent edition of this publication.

Disclaimer
This material has been published on behalf of Edexcel and offers high-quality support for the delivery of Edexcel qualifications. This does not mean that the material is essential to achieve any Edexcel qualification, nor does it mean that it is the only suitable material available to support any Edexcel qualification. Edexcel material will not be used verbatim in setting any Edexcel examination or assessment. Any resource lists produced by Edexcel shall include this and other appropriate resources.

Copies of official specifications for all Edexcel qualifications may be found on the Edexcel website: www.edexcel.com

Contents

Introduction

About Functional Skills Mathematics

Congratulations! If you are studying this book you are working towards the Edexcel Functional Skills qualification in Mathematics at Level 2.

Functional Skills are designed to give you the skills you need to be confident, effective and independent in education, work and everyday life.

In Functional Skills Mathematics, you will work on questions set in real life that will help you develop the skills you will need in life.

22 Three major cities in the UK were surveyed to find out about people's exercise habits.

The results of the survey claim that:
- three out of every 100 men do enough exercise
- one out of every 50 women do enough exercise

Q a) Work out the percentage of men and the percentage of women that the survey claims do enough exercise.

You will learn how to apply the mathematics you know to solve real-life problems and you will develop these important process skills:

Process skill	What it is	What it means	% of marks in the exam
Representing	Selecting the mathematics and information to model a situation	You decide how to tackle the problem you have been given	30–40%
Analysing	Processing and using mathematics	You apply your mathematics skills and understanding to solve the problem	30–40%
Interpreting	Interpreting and communicating the results of the analysis	You interpret the problem and make conclusions, justifying your response	30–40%

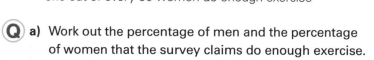

About this book

Edexcel Functional Skills Mathematics is specially written to help you achieve the Edexcel Functional Skills qualification in Mathematics at Level 2.

Eleven chapters based on familiar maths topics such as number and measures. Enables you to relate Functional Skills mathematics work to your GCSE work

examzone pages show you sample student answers with tips and reports on real exam questions

Apply what you have learnt by trying the Assessment practice questions that come after each group of chapters

The final Exam-style practice section helps prepare you for real exam questions

Check your answers to the shorter 'Let's get started' questions to see that you are on the right track

Contents

The features of the chapters

Each chapter begins with a KnowZone section:

Recap the key maths knowledge and understanding on the topic

Clear objectives tell you what you will learn in the section

Worked examples show you how to solve functional-style questions

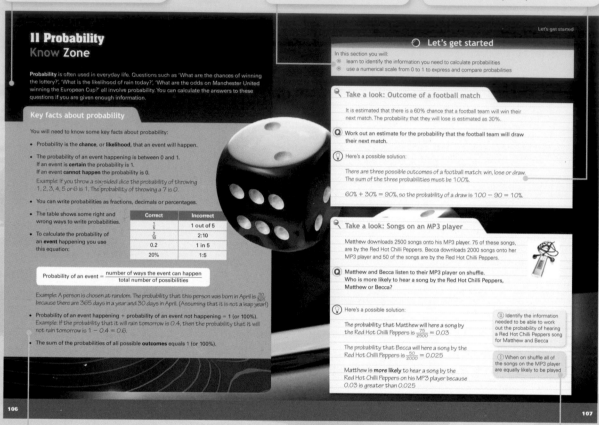

Brief examples show the key skills you need to be able to use

See exactly where the process skills are being used:
- **R** = representing
- **A** = analysing
- **I** = interpreting

After KnowZone, three sections provide progressively more challenging questions for you to practise:

Let's get started short, straightforward questions to get you started

We're on the way more challenging questions with some of the features of real functional maths questions

Exam ready! longer and more complex questions to help prepare you for the exam

Have a go

1   A poll showed the probability that Labour will win the next general election as 34%.

(Q) **What is the probability that Labour will *not* win the next general election?**

2   A survey into trains being on time showed that the probability the London to Manchester train arrives on time is 0.779

Plenty of practice questions for you to try

---

have to wait more than 10 minutes.

The medical centre has a target.
Each day, the number of people who have to wait more than 20 minutes should be less than 20% of the total number of appointments.
The centre manager says that the medical centre meets this target.

(Q) **b) Do the figures for the week of the survey support this claim?**

ResultsPlus
Exam Tip

It is important that you set out your working clearly to show the methods you have used to arrive at your solution.

This gives exam advice, useful hints and warns you of common mistakes that students frequently make

---

(Q) What is the probability that Cheadle Town only wins one game?

11   A couple plan to have three children. Assume that the probability of a baby being born a boy or girl is the same.

(Q) Identify the possible outcomes for the gender of the children. Calculate the probability that the three children will have the same gender.

Think First!

Consider the order of the birth of the children.

Use the Think first boxes for tips to help you get started on a question

---

$R = \dfrac{dat}{h}$

R = R-value
d = temperature difference (Fahrenheit)
a = area (square feet)
t = time (hours)
h = heat loss (Btu)

Insulation materials	R-value for 1 inch
Loose fill	2.08
Foam	5.3
Blown loose fill	3.6

Fact
Btu stands for British thermal unit.

Facts, abbreviations and specialist terms explained

---

this claim?

Now you can:

- Use a numerical scale to express and compare probabilities
- Identify a range of possible outcomes of combined events and record the information in tree diagrams or tables
- Make a model of a situation to work out probabilities

Check your progress using this summary box at the end of the chapter

About ActiveTeach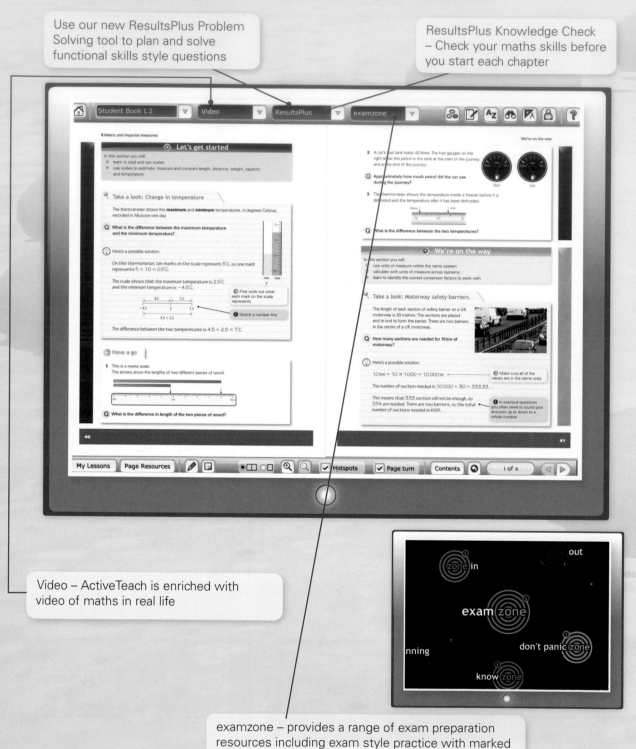

Use ActiveTeach to view the course on screen with exciting interactive content.

Use our new ResultsPlus Problem Solving tool to plan and solve functional skills style questions

ResultsPlus Knowledge Check – Check your maths skills before you start each chapter

Video – ActiveTeach is enriched with video of maths in real life

examzone – provides a range of exam preparation resources including exam style practice with marked student answers and examiner feedback

The Functional Skills Mathematics assessment

You will take one paper which will be marked by Edexcel. The exam is 'pass' or 'fail' – there are no other grades.

There are many opportunities a year to sit the exam and you can take the exam as many times as you and your centre wish.

The exam lasts 1 hour 30 minutes and you can use a calculator throughout.

The exam paper has three sections. Each section has a theme, such as jobs or the garden.

Section 1 – first theme	16 marks
Section 2 – second theme	16 marks
Section 3 – third theme	16 marks
Total	**48 marks altogether**

You must practise showing your working at all times.

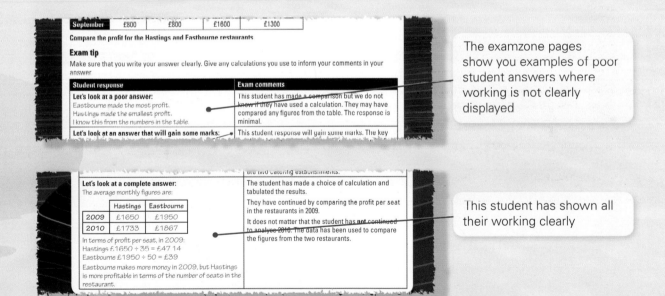

The examzone pages show you examples of poor student answers where working is not clearly displayed

This student has shown all their working clearly

1 Number
Know Zone

In this chapter you will find situations that need a functional approach to applying number skills. Think through how you will begin to solve each problem before you apply your number skills.

Read, write, order and compare positive and negative numbers

A hospital needs to have more staff working in Accident and Emergency when the temperature drops below −2°C.
The recorded temperatures (in °C) over the last ten days were 4, 2, 0, −2, −3, −5, −7, −6, 0 and −4.

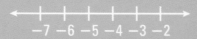

Over the last ten days the hospital needed more staff for five days.

Remember that −3, −5, −7, −6 and −4 are smaller than −2

Understand the meaning of negative numbers in a practical context

Football league tables have a column for goal difference.
Portsmouth FC had a goal difference of −17 on 5 February 2010.
Explain the meaning of the statement 'goal difference of −17'.
Portsmouth had 17 more goals scored against them than they scored against other teams.

Carry out calculations with numbers of any size in practical contexts

You will need to use the correct level of accuracy when you solve a problem.
Three students are sharing the cost of a shopping bill equally. The bill comes to £38.26
How much money should each student pay towards the shopping bill?

£38.26 ÷ 3 = £12.7533333
Each student should contribute £12.76
There will be 2p left over!

A decision has to be made to **round up** the money – if we round down, the bill will not be paid.
£12.75 × 3 = £38.25

In this section you will:
- ◉ use a calculator to represent situations efficiently
- ◉ use prime numbers
- ◉ check your answers to see if they work

🔍 Take a look: Bar codes

Bar codes have numbers printed at the bottom.
This bar code number consists of:
- a single number (5)
- two larger numbers made up of three two-digit numbers between 01 and 20 (01 10 13 and 10 01 18)

5   011013   100118

(Q) Write a bar code number that consists of prime numbers between 1 and 20, similar to the example above.

> **Fact**
>
> A **prime number** is a **whole number** that can only be divided exactly by itself and one. The number 4 is not a prime number because it can also be divided by 2.

(💡) Here's a possible solution:

List the **prime numbers** between 1 and 20:
2, 3, 5, 7, 11, 13, 17, 19

> **R** The number 1 is not a prime number!

A suitable bar code would be:
3 021911 070702

🔍 Take a look: Bill

A plumber sends a bill to a customer.
It contains the following details.

Replace faulty radiators:	7 hours labour at £40 per hour	£280
Travel:	6 miles at 40p per mile	£24
	Total =	£520

There are some errors in the bill. What is the correct total?

> **A I** Two of the calculations in the bill are incorrect — you need to be able to check the calculations and communicate any errors you find

(💡) Here's a possible solution:

6 miles at 40p per mile should be £2.40
£280 + £2.40 is £282.40

The correct total should be £282.40

Have a go

1   A football club's ground has an area where supporters have a meal before or after the match. The prices of these tickets for a match are given in the table below.

Ticket type	Price
Pre-match meal	£112
Post-match meal	£135

The football club sold 230 pre-match meal tickets and 290 post-match meal tickets.

Q What is the total cost of the meal tickets sold?
Write your answer to the nearest £100.

2   A scientist recorded where certain types of fish are found in the ocean.

Fish	Where they are found (in m below sea level)
Skate	between 30 m and 50 m
Herring	not lower than 25 m
Angler fish	between 60 m and 100 m

Q a) What is the minimum distance between where Skate and Angler fish are found?

b) What is the minimum distance between where Herring and Skate are found?

3   Colin, Gaby, Anoop and Petruk play table tennis. Each person plays each of the others once only.
   - Colin beat Gaby
   - Anoop lost all of his matches
   - Petruk won all of his matches

They decide to award 1 point for a win, 0 points for a draw and −1 point for a loss.

Think First!

You will need to:
   - Write down the games that were played
   - Interpret the results.

Q Construct a results table for the table tennis games. Show the results of the matches and the number of points each person got.

Think First!

You need to decide:
   - How to construct the table
   - What information to include.

We're on the way

In this section you will:
- use different techniques to inform your decision making
- interpret situations
- give a level of accuracy of results
- interpret and analyse unfamiliar situations

Take a look: Temperatures in Iceland

This table shows the **maximum** and **minimum** temperatures, in °C, for two Icelandic towns for November to April.

Month		Nov	Dec	Jan	Feb	Mar	Apr
Reykjavik	Max	3.4	2.2	1.9	2.8	3.2	5.7
	Min	−1.3	−2.8	−3.0	−2.1	−2.0	−0.4
Akureyri	Max	2.6	1.3	0.9	1.7	2.1	5.4
	Min	−3.5	−5.1	−5.5	−4.7	−4.2	−1.5

Q Compare the maximum and minimum temperatures in the two towns.

R The question allows for an open response – the decisions on what techniques to use are your own

Here's a possible solution:

The difference between the minimum and maximum temperatures for each town and for each month could be found:

The temperature difference in Reykjavik for November is 4.7°C
The temperature difference in Akureyri for November is 6.1°C

	Temperature difference					
Month	Nov	Dec	Jan	Feb	Mar	Apr
Reykjavik	4.7	5.0	4.9	4.9	5.2	6.1
Akureyri	6.1	6.4	6.4	6.4	6.3	6.9

A You could find the temperature difference in Reykjavik for November by subtracting the minimum temperature from the maximum temperature:
$3.4 − (−1.3) = 4.7$

The mean maximum and minimum temperatures for both towns could be compared. For example, the mean maximum temperature for Reykjavik is 3.2°C:

$$(3.4 + 2.2 + 1.9 + 2.8 + 3.2 + 5.7) ÷ 6 = 3.2°C$$

	Mean temperatures	
Reykjavik	Max	3.2
	Min	−1.9
Akureyri	Max	2.3
	Min	−4.1

The mean maximum and minimum temperatures suggest that Akureyri has lower temperatures than Reykjavik.
The temperature differences suggest that Akureyri has a greater variation in temperature than Reykjavik.

A You need to communicate your findings clearly

Have a go

4 Four doctors work at an overnight surgery.

The rate of pay is £85 per hour.
The surgery is open from 1900−0800 each day.
Two doctors need to be on duty at any time.

(Q) **What will the bill for doctors' wages be for August?**

5 A shopkeeper buys items.
She can choose where to buy them.
The shopkeeper needs 360 items.

	Supplier A	Supplier B
	Box of 24	Box of 36
Cost per box	£21.50	£26.50
Cost of delivery (per order)	£9.85	£28.75

Think First!

The delivery charge is per order, not per box.

(Q) **Compare the cost of buying 360 items from each supplier.**

⊙ Exam ready!

In this section you will:
◉ work with problems that require coordination of different features

🔍 Take a look: Gritting the roads

A county council has to estimate the cost of grit needed for winter.

The council spreads grit on the roads when the temperature drops below 0 °C.

The council has the following predictions of temperatures below 0 °C for next winter.

Temperature (°C)	−1	−2	−3	−4	−5	−6 or below
Number of days	5	5	4	3	3	4

As the temperature decreases, the amount of grit spread on the roads increases.

Temperature (°C)	−1	−2	−3	−4	−5	−6 or below
Grit (tonnes)	250	275	300	350	375	450

The grit is made up of sand and salt in the ratio of 4:1 by weight.
The costs are fixed.
- Sand is £12.20 per tonne
- Salt is £71.00 per tonne

Q **a)** Find the amount of grit needed.
b) Estimate the cost of grit for next winter.

Here's a possible solution:

I You need to find the total **amount** of grit needed for winter

a) Amount of grit needed
$$= (5 \times 250) + (5 \times 275) + (4 \times 300)$$
$$+ (3 \times 350) + (3 \times 375) + (4 \times 450)$$
$$= 7800 \text{ tonnes}$$

R **A** You need to work out how much of the grit is sand and how much is salt

b) The ratio of sand to salt is 4:1 (four parts sand to one part salt).
Sand will make up $\frac{4}{5}$ of the grit. Salt will make up $\frac{1}{5}$.

The amount of sand will be $\frac{4}{5} \times 7800 = 6240$ tonnes.

The amount of salt will be $\frac{1}{5} \times 7800 = 1560$ tonnes.

The estimated cost of the grit is:
$$(12.20 \times 6240) + (71 \times 1560) = 76\,128 + 110\,760$$
$$= £186\,888$$
$$= £187\,000 \text{ (to the nearest £1000)}$$

Have a go

6 A company needs to hire a team of IT specialists for two months. There must be at least one consultant, two programmers and two technicians in the team.

The company has a budget of £75 000. They intend to spend as much of this budget as possible.

A recruitment agency quotes the following costs for IT specialists.

IT specialist	Cost for two months
Consultant	£18 800
Programmer	£12 000
Technician	£7000

Q Choose the team of IT specialists for the company to hire.

7   A charity sent a questionnaire to its supporters. The table below gives information about the response to one question.

strongly agree	2363
agree	2357
neither agree nor disagree	12 455
disagree	24 167
strongly disagree	234

(Q) Analyse the results for this question and advise the charity on how they should communicate the overall result for this question.

8   A cathedral choir needs three boys and two girls to become choristers.

Choristers become students at the Cathedral School.

The Cathedral School will pay 50% of the school fees for each chorister.

The school fees are given in the table below.

Student type	Fee per year
Day	£10 081
Boarder	£17 674

(Q) a) Work out how much the Cathedral School will pay for a day student and for a boarder.

The Cathedral School held singing tests for selected boys and girls. Each boy and girl was judged on a five point scale.

1 − outstanding   2 − very good   3 − good   4 − satisfactory
5 − room for improvement

Boys			Girls		
Name	Scale	Student type	Name	Scale	Student type
Arun	3	day	Alice	4	day
James	5	boarder	Bernice	3	boarder
Rohan	2	boarder	Cherry	3	day
Soha	2	boarder	Jasmin	2	day
Steven	3	day	Uzma	1	boarder
Thomas	1	day			

The Cathedral School has £35 000 to spend on fees for choristers.

(Q) b) Can the Cathedral School afford to pay 50% of the school fees for all the students who were judged outstanding or very good?

9   Sam uses rainwater to flush her toilets and wash her clothes.

Sam has estimated that she can collect water from a rectangular area of $1\,188\,000\,cm^2$.

In January, there was 68 mm of rainfall.

(Q) a) Work out the volume of rainwater Sam collected in January.
Give your answer in litres.

Sam estimates that she needs 2100 litres of water per week to flush her toilets and wash her clothes.

> **Fact**
> 1 litre = $1000\,cm^3$

(Q) b) How many weeks will the water she collected in January last?

Now you can:

- Use negative numbers in situations
- Use numbers to interpret problems
- Communicate using numbers

Fractions

- Write fractions in their simplest form.
  Do this by dividing the numerator and denominator by their highest common factor (HCF).

  Example: $\frac{24}{36}$ in its simplest form is $\frac{2}{3}$. You divide **both** the numerator and the denominator by 12.

- Equivalent fractions are fractions with an identical simplest form.

  Example: $\frac{14}{35}$ and $\frac{8}{20}$ **both** have the simplest form of $\frac{2}{5}$, so they are equivalent.

- A mixed number has a whole number part and a fraction part.

  Example: $2\frac{3}{5}$

- You often need to express one quantity as a fraction of another quantity. You might need to round quantities first.

  Example: 7 as a fraction of 49 is $\frac{7}{49} = \frac{1}{7}$

  Example: 499 as a fraction of 753 is roughly $\frac{500}{750} = \frac{2}{3}$.

- You often need to find fractions of quantities in everyday situations.

  Example: three-sevenths of a population of 28 000 voted Labour:

  $\frac{3}{7} \times 28\,000 = 12\,000$ or $\frac{3 \times 28\,000}{7} = 3 \times 400 = 12\,000$

Decimals

- You need to know how to order and compare decimals.
  Compare place values digit by digit.

  Example: 3.0203 is more than 3.015 because 2 hundredths is more than 1 hundredth.

- You need to feel confident about calculating with decimals in practical situations.

  Example: £1 = € 0.88.

  To convert pounds to euros, multiply the amount in pounds by 0.88

  To convert euros to pounds, divide the amount in euros by 0.88

- You need to be able to round numbers appropriately.

  Example: the number 24.9058 is 24.9 to 1 decimal place and 24.91 to 2 decimal places.

Percentages

- You need to be able to convert between equivalent fractions, decimals and percentages.
  Example: $0.7 = \frac{7}{10} = 70\%$
  $0.666\ldots = \frac{2}{3} = 66.6\ldots\%$

- It is useful to use these equivalences when you order fractions, decimals and percentages.
  Example: $\frac{3}{4} = 0.75$, $80\% = 0.8$, $\frac{2}{3} = 0.666\ldots$.
  Converting to decimals, you can easily see that you can arrange the numbers in
  ascending order as $\frac{2}{3}$, $\frac{3}{4}$, 80%.

- You need to be able to find a percentage of a quantity.
  To find the VAT (at 17.5%) that you pay on something that costs £25,
  multiply the cost by 0.175
  £25 × 0.175 = £4.38 (rounding to the nearest penny)

- Percentage changes occur in practical problems.
  For example, prices can by increased by a percentage, such as VAT, or
  decreased by a percentage, such as a sales discount.
  To increase a quantity by 5%, a quick method is to multiply the quantity
  by 1.05 (1 + 0.05).
  To decrease a quantity by 8%, the quickest way is to multiply the
  quantity by 0.92 (1 − 0.08).

- You need to know how to write one number as a percentage of another.
  Example: to write 158 as a percentage of 795:
  158 ÷ 795 × 100 = 19.87%
  Therefore 158 is roughly 20% of 795.

Finding the percentage change

- To calculate percentage change use this formula:
  You buy something for £400. You sell it for £300.
  What is the decrease in percentage price?

$$\text{Percentage change} = \frac{\text{change}}{\text{original value}} \times 100$$

'Change' is the amount of increase (profit) or decrease (loss). Here the change is £100 decrease (loss).

$\frac{100}{400} \times 100 = 25\%$

Let's get started

In this section you will:
- find fractions and percentages of quantities in practical problems
- add, subtract, multiply or divide with decimals in practical problems
- round to an appropriate number of decimal places in practical problems
- increase or decrease quantities using percentages in practical problems
- use decimal, fraction and percentage equivalences to find solutions to practical problems

Take a look: Weight loss

Chris weighs 77 kg. She wants to lose 11 kg.

Q What is 11 kg as a percentage of Chris's current body weight?

💡 Here's a possible solution:

$\frac{11}{77}$

R First write the weight loss Chris wants as a fraction of her actual weight

$\frac{11}{77} = 0.1429$ to 4 decimal places

$0.1429 \times 100 = 14.29\%$

A Convert this fraction to a decimal. Then multiply by 100 to find 11 kg as a percentage of Chris's actual weight

11 kg is 14.29% of Chris's actual body weight.

Take a look: Managing on a budget

Rachel's monthly income is £1350. She spends a third of her income on bills. She spends $\frac{2}{5}$ of her income on her mortgage.

Q How much money does Rachel have left each month?

💡 Here's a possible solution:

£1350 ÷ 3 = £450
Rachel spends £450 on bills.

A First find out the costs of Rachel's mortgage and bills

£1350 × 2 ÷ 5 = £540
Rachel spends £540 on her mortgage.

A To find a fraction of an amount, multiply by the numerator and divide by the denominator

1350 − 450 − 540 = 360
Rachel has £360 left.

A Work out how much money is left

Have a go

1 Oscar ate $\frac{1}{8}$ of a pizza.
His Auntie Cathy ate three times as much pizza.
They left the rest of the pizza for Uncle Simon.

(Q) What fraction of the pizza did they leave for Uncle Simon?

2 Abi bought a second-hand guitar for £500.
She sold the guitar on an auction website for £375.

> **Think First!**
> First work out the amount Abi loses.

(Q) What is the loss as a percentage of the purchase price for Abi's guitar?

3 Helen did a sponsored climb of Mount Kilimanjaro for charity.
She got £6527 of sponsor money.
The charity claims that $\frac{2}{3}$ of sponsor money is spent on good causes. The rest of the money pays admin costs.

(Q) Work out how much the admin costs are.

Take a look: Converting height

Divya is filling in a form.
She needs to enter her height in metres.
She knows that her height is 5 feet 4 inches.

> **Fact**
> 12 inches = 1 foot
> 1 inch = 2.54 cm

(Q) Work out Divya's height in metres.

Here's a possible solution:

5×12 inches = 60 inches
Divya is 5 foot 4 inches. Her height in inches = 64 inches.

> **A** Convert 5 feet 4 inches into inches

$64 \times 2.54 = 162.56$ cm

> **A** Convert inches to cm

162.56 cm $= 1.6256$ m $= 1.63$ m

> **R** Round to the nearest centimetre

Divya's height is 1.63 m.

🔍 Take a look: Pay increase

Jim's salary is £32 000 per year.
He pays tax of 20% on everything he earns above £6500

(Q) (a) How much tax does Jim pay per year?

Jim is going to get a pay rise of 5%.

(Q) (b) How much money will Jim take home each month after the pay rise?

💡 Here's a possible solution:

a) $32\,000 - 6500 = 25\,500$
$0.2 \times 25\,500 = £5100$

Jim pays £5100 tax each year.

b) $0.05 \times 32\,000 = 1600$
$32\,000 + 1600 = 33\,600$

$33\,600 - 6500 = 27\,100$
$0.2 \times 27\,100 = 5420$
$33\,600 - 5420 = 28\,180$
$28\,180 \div 12 = 2348.333\ldots$

Jim takes home £2348.33 each month.

R First work out how much money Jim pay's tax on

A Calculate 20% of 25 500

A First work out Jim's new salary

A Work out how much money Jim has each month after paying tax

I The answer is a sum of money. Give your answer to two decimal places

⊕ Have a go

4 Here is a question in a survey: "What proportion of your monthly income do you save?"

Here are the responses of three people:

Sam — About 10%

Jon — £8 out of every £100 I earn

Ella — An eighth

Think First!

Decide if you want to compare the responses using percentages or fractions.

(Q) Who saves the smallest proportion of income?

5 David wants to know his weight in kilograms.
David weighs 17 stone 3 pounds.

(Q) Work out David's weight in kilograms.

Fact

1 stone = 14 pounds
1 kg ≈ 2.2 lbs

6 George used 6453 kWh of gas one year.
George has received a bill for £318.09
Here is the information given on his gas bill.

Gas unit rates	7.35p per kWh of gas for up to 2680 kWh of gas used per annum	3.21p per kWh of gas used above 2680 kWh per annum

Think First!

Note that there are two different rates per kWh. The rate depends on how much gas is used. You do **not** need to understand the meaning of kWh to solve the problem.

(Q) Check if the bill has been calculated correctly.

7 Rashid needs to buy car insurance. The normal price of his annual insurance premium is £1035. Rashid can get a 30% discount. If Rashid buys his car insurance online, he gets another 5% discount off the price.

Think First!

Is a 30% discount followed by a further 5% discount the same as a 35% discount?

(Q) Rashid decides to buy his car insurance online.
How much will he pay for his car insurance?

8 Maisie buys twelve cans of drink for £4.80
She will sell the drinks from her ice cream van.

Maisie wants to make a 15% profit on each can.
She decides to round the selling price to the nearest ten pence.

(Q) How much should she sell each can for?

9 Lekita trains every week for a swimming competition.
In week 1 Lekita swims 500 m.
She wants to swim 20% further each week than she swam the previous week.
The pool she trains in has a length of 50 m.

(Q) Work out the distance and number of complete lengths Lekita should swim in weeks 2, 3 and 4.

10 Margaret decides to sell some of her gold and silver.
She sends 55 g of gold and 36 g of sterling silver to a company called 'Cash for Gold!'.

Here are their rates:

(Q) How much should the company pay Margaret?

CA$H FOR GOLD
We buy gold for 845.6p per gram
We buy sterling silver for 420.5p per gram

We're on the way

In this section you will:
- use proportion in practical problems
- use exchange rates in practical problems
- decide if it is appropriate to multiply or divide with decimals in practical problems
- express one quantity as an approximate fraction or percentage of another quantity, as a strategy to solve a practical problem

Take a look: Changing exchange rates

In 2006 Liz and Dave went to Barcelona. They spent €1500 in total. At the time €200 was equivalent to £140.
Liz and Dave plan to go to Barcelona again this year.
They decide to spend the same amount, about €1500.
Liz finds this exchange rate: €1 = £0.877 140 73 GBP

Q The trip will cost Liz and Dave more this year than in 2006. How much more?

A Find out how many lots of €200 are equal to €1500. Then work out the cost, in pounds, that they spent in 2006

A Find the value of 1500 euros in pounds

I Interpret your answer

Here's a possible solution:

4 years ago 1500 ÷ 200 = 7.5
7.5 × £140 = £1050

0.877 140 73 rounds to 0.88
1500 × 0.88 = £1320
Extra cost is 1320 − 1050 = £270

The trip this year will cost Liz and Dave approximately £270 more.

Take a look: Making cakes

Julie makes cakes for a local shop. She makes five cakes each week. There are 500 g of flour in each cake.

Julie usually spends £15 each week on ingredients.
She spends $\frac{1}{3}$ of the £15 on flour.
The shop owner pays Julie £5.50 for each cake.

Q One week there is a special offer. Work out how this special offer affects Julie's weekly percentage profit.

Special offer
Self-raising flour
1 kg and 2 kg bags
Reduced by 50%!

Here's a possible solution:

£15 ÷ 3 = £5
A 50% price reduction means Julie saves
£2.50 when she buys this week's ingredients.

R Extract the relevant information from the question – the weight of ingredients is information you do not need

A Find both the normal cost and the reduced cost of flour

Normal profit = 5 × £5.50 = £27.50 − £15 = £12.50
Normal percentage profit = 12.50 ÷ 15 × 100 = 83.3%

A To find percentage profit, calculate the profit first. Then divide the profit by the cost price

New profit = £27.50 − £12.50 = £15
New percentage profit = 15 ÷ 12.50 × 100 − 120%

A Find the change in percentage profit

Julie's percentage profit increases by approximately 37%.

I Explain your findings

Have a go

11   A supermarket sells fish fingers in two sizes of box.

Regular size	Box of 8	£3.99
Jumbo size	Box of 16	£5.99

The supermarket has a special offer.
Mrs Nahit needs 50 fish fingers.

Today only!
Buy any box of fish fingers –
get a second box half price!
(cheapest box is half price)

Q What is the cheapest way to buy fish fingers?

12   Mike buys 24 keyrings for £9.20
He sells the keyrings to the members of the maths club for 60p each.

The price of packs of 24 keyrings increases to £10.60
Mike needs to decide how much to sell each keyring for now.
He wants to keep the same percentage profit. He does **not** want to
charge a price where he has to give change using 1p or 2p coins.

Q How much should Mike sell each keyring for now?

13 An author gets royalty cheques once every six months.
Her contract states that she earns 8.4% in royalties on all her books sold.

Here is a record of how much
she was paid in 2009.

Date	March 2009	September 2009
Amount paid	£4982.67	£3240.56

The author wants to work out how
much the sales of her book came to in 2009.

Q Work out the approximate total value of book sales

Take a look: Budgeting

Bronwen spends 40% of her weekly income on bills. She spends $\frac{1}{6}$ of her weekly income on food. She has about £130 left to spend on other things.
Bronwen cannot buy all the things she wants with only £130. She decides to halve the amount she spends on food. She thinks this will double the amount she has to spend on other things.

Q **a)** Work out Browen's total weekly income.

b) Bronwen halves the amount she spends on food. Work out how much Bronwen has left to spend on other things.

c) Is it double the amount she had before?

Here's a possible solution:

a) $\frac{1}{6}$ is roughly 17%

A Work out the percentage of her income Bronwen spends on food

40% + 17% = 57%

A Work out the percentage of her income Bronwen spends on bills and food together

100% − 57% = 43%
43% of the total budget = £130

A Work out what percentage of Bronwen's income the £130 is

£130 ÷ 43 × 100 is roughly £300

A Work out the value of Bronwen's total weekly income

b) $\frac{1}{2} \times \frac{1}{6} = \frac{1}{12}$

R Bronwen plans to halve what she spends on food. Work out how much she will spend on food now

$\frac{1}{12}$ of £300 is £25

She spends 40% of £300 on bills:
40% of £300 = £120
£300 − £120 − £25 = £155

A Work out how much money Bronwen will have left when she has paid for food and bills

Bronwen has £155 left to spend on other things.

c) This is not double £130.

I Evaluate your findings

 **Have a go**

14 A charity plans a coffee morning every year.
The table shows the sales figures for last year's coffee morning.

Entrance tickets	Refreshment sales	Stalls and raffle
£28	£115	£108

- The charity expects the number of people coming to the coffee morning this year to increase by $\frac{1}{3}$.
- They expect the money from the sale of refreshments, from the stalls and the raffle to be the same.
- They plan to increase the entrance ticket fee by 25%.

(Q) Work out how much the charity expects to get from this year's coffee morning.

15 A hotel wants at least $\frac{3}{5}$ of its rooms to be occupied at weekends.
The table gives information about the rooms occupied last weekend.

	Single rooms	Twin rooms	Double rooms
Number of rooms in total	73	176	224
Number of rooms occupied	53	97	143

(Q) Use the information in the table to find if at least $\frac{3}{5}$ of the rooms were occupied last weekend.

16 Matt thinks the roots of a tree are too close to the foundations of his house. He finds this information on the Internet:

- This type of tree can grow to 18 m in height
- The roots spread horizontally below ground. The length of the roots can be $\frac{3}{8}$ of the height of the tree

If the roots of the tree are likely to be within a metre of the foundations of his house, Matt will chop the tree down.

Here is a diagram showing the tree's position in relation to Matt's house.

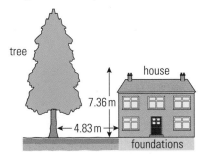

(Q) Should Matt chop the tree down?

17 Dunford School wants to promote healthy eating.

The school secretary writes a report on the eating habits
of students at lunchtime.
Here is part of the report:

> "On average 65% of students eat school dinners
> every day. About $\frac{2}{3}$ of the 65% of students have
> chips. Four out of every five students having
> chips at lunchtime also often have a high-calorie
> dessert."

A national report states that at least 30% of all students eat chips and a
high-calorie dessert every day for lunch.
There are 1200 students at Dunford school.

Q Compare the lunchtime eating patterns of students at Dunford
School with the national report's statement.

18 Ben is going to cook chicken stew for 45 people.

The recipe needs three small onions per person.
The onions are sold in 0.25 kg bags.
There are 8–10 onions in each bag.
The onions cost £2.39 a kilogram.

Q **a)** Advise Ben on how many bags of onions to buy.
b) Work out the cost of the onions.

Exam ready!

In this section you will:
- use a range of calculation techniques using fractions, decimals and percentages
- interpret and communicate solutions to multi-stage practical problems, and use calculations to justify any statements made

Take a look: Owning a restaurant

The owner of a restaurant wants at least $\frac{2}{3}$ of all meals to be two or more courses.
The table shows the meals ordered in the restaurant one Sunday lunchtime.

	Number of meals
Starter only	5
Starter and main course	45
Main course only	34
Main course and dessert	43
Starter, main course and dessert	54

Q Work out if at least $\frac{2}{3}$ of the meals were at least two courses.

Here's a possible solution:

Total number of two and three course meals = 45 + 43 + 54 = 142
Total number of meals = 181

> **A R** Find out how many meals there were of two or more courses. Find the total number of meals

142 ÷ 181 = 0.7845... = 78.45%

The restaurant owner wanted at least $\frac{2}{3}$ of meals to be two courses or more.

$\frac{2}{3}$ = 0.6666... = 66.67%

Therefore, at least $\frac{2}{3}$ of the meals were 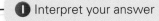 for two or more courses.

> **I** Interpret your answer

⊙ Have a go

19 Mark owns a furniture shop. He buys furniture from furniture makers.
He sells the furniture in his shop.
The table shows his sales for January.

	Item	Purchase cost (£)	Sales (£)
January	beds	3100	4540
	wooden furniture	6980	8000
	sofas	9020	12 600

Q Work out his total percentage profit for January.

20 The table shows the sales figures for jeans at a clothes shop in March and April.

Type	March	April
Bootleg	84	69
Skinny	73	122

Q **a)** Work out the percentage decrease in sales of bootleg
jeans from March to April.

The shopkeeper thinks that sales from April to May will follow
the percentage increase or decrease they did from March to April.

Q **b)** How many of each type of jean should the shopkeeper order
for May?

21 Ali used a questionnaire for a GCSE maths project.
She wants to display the results of the survey on a poster.
She decides to use percentage bar charts to show the answers to each question.
Each percentage bar chart will be exactly 10 cm long.

For one question, $\frac{1}{2}$ the people said Yes, $\frac{1}{4}$ of them said No and $\frac{1}{4}$ said Don't know.

Ali showed the answers to some of her questions in a diagram:

50%	25%	25%

←——— 10 cm ———→

Key:
■ Answered Yes   □ Answered Don't know   ■ Answered No

Here are the answers to another question.

Yes	33 people
Don't know	15 people
No	12 people

Q Draw an accurate percentage bar chart like the one above for
the answers to this question.

22 Three major cities in the UK were surveyed to find out about people's exercise habits.

The results of the survey claim that:
- three out of every 100 men do enough exercise
- one out of every 50 women do enough exercise

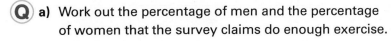

Q a) Work out the percentage of men and the percentage of women that the survey claims do enough exercise.

'Enough exercise' means exercise for more than 20 minutes 5 times a week.

The table shows information about responses to one question in the survey:

"Do you exercise for more than 20 minutes 5 times a week?"

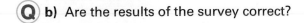

	Glasgow		Leeds		London	
Number of surveys given out	1000		1500		2500	
Number of responses	women	men	women	men	women	men
	456	382	699	582	1076	860
Number of 'Yes' responses	22	4	23	10	36	37

Q b) Are the results of the survey correct?

Now you can:

- Find fractions or percentages of quantities to solve practical problems
- Use decimal, fraction and percentage equivalences appropriately to solve a problem
- Use proportions to compare trends and to predict behaviour
- Evaluate answers that use decimals, fractions and percentages in the context of the question in order to make conclusions

3 Ratio and proportion
Know Zone

Ratios and proportions are very useful tools for everday problem solving.

Ratio

- A **ratio** compares two or more quantities.
  Example: There are seven men and nine women in a group, so the ratio of men to women is 7:9

- Ratios are usually written in their simplest form.
  To do this, divide both sides by the same common factor.
  Example: You can write the ratio 25:45 as 5:9 if you divide **both** 25 and 45 by 5.

- The order of the numbers in a ratio is important.
  Example: 5:9 is not the same as 9:5

- Ratios do **not** have units.
  Example: You can write the lengths 5 m and 75 cm
  in ratio form as 500:75
  In its simplest form 500:75 is 20:3

- The scale on a plan or map is usually written as
  a ratio.
  Example: A scale of 1:25 000, means 1 cm
  on a map represents a distance of 25 000 cm
  (250 m) or, as shown on this map, 4 cm on the
  map represents 1 km.

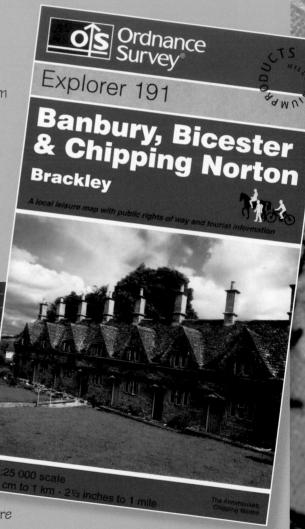

Proportion

- When there is a ratio between quantities,
  the quantities are in the same **proportions**.
  Example: A ratio 3:5 has a total of eight
  parts because $3 + 5 = 8$. So, the
  proportions are $\frac{3}{8}$ and $\frac{5}{8}$

- At Level 2 you will often have to
  approximate ratios and proportions.
  Example: In the ratio 248:743 the proportions are
  roughly $\frac{1}{4}$ and $\frac{3}{4}$

In this section you will:
- find and express the numbers you need as ratios in their simplest form
- divide quantities into a given ratio
- make decisions using proportion

Take a look: Pocket money

Mrs Fox is going to share £15 pocket money between her two children, Charlotte and Hannah.

She will share the £15 according to how much time each child works doing jobs around the house.

Charlotte works for 40 minutes.
Hannah works for 60 minutes.

(Q) **How much pocket money does each child get?**

Here's a possible solution:

$40:60 = 2:3$

$2 + 3 = 5$

This means that Charlotte gets $\frac{2}{5}$ of the £15
Hannah gets $\frac{3}{5}$ of the £15

Charlotte gets $\frac{2}{5} \times £15 = £6$.
Hannah gets $\frac{3}{5} \times £15 = £9$.

Here's another possible solution:

$40:60 = 2:3$
$2 + 3 = 5$

$£15 \div 5 = £3$

Charlotte gets $2 \times £3 = £6$.
Hannah gets $3 \times £3 = £9$.

R Find the numbers you need to express as a ratio

A Simplify the ratio if you can

A Add the number of shares to find the total

A Multiply each proportion by the total amount of pocket money to find out how much each child gets

A Write the times as a ratio in its simplest form and find the total number of shares

A Find the value of one share

A Multiply one share by the number of shares for each child to see how much pocket money they get

Have a go

1 Shelley plans a party for 60 children.

She is going to make a fruit cocktail out of orange juice and raspberry juice in the ratio 5:3

Shelley will make enough cocktail so that each child can drink two glasses.

She will put approximately $\frac{1}{5}$ of a litre of cocktail in each glass.

(Q) How many litre cartons of orange juice does Shelley need to buy?

2 Screen wash concentrate for cars is sold in 5-litre bottles.

Samil buys two bottles.
He estimates this will be enough to last him through the winter.
He knows that he fills the screenwash container in his car three times in winter.

The instructions on each bottle of concentrate say:

> Mix 3 parts concentrate with
> 5 parts water.

The screenwash container in Samil's car has a capacity of 8 litres.

(Q) Decide whether Samil's estimate, that two bottles of concentrate is enough to last him through winter, is correct or not.

3 A youth worker is going to do a survey of three community groups. He decides to survey 30 people.

He will choose people at random from each group. He wants to survey the number of people from each group in the same proportion as the size of the group.

Here is some information about the groups:

Ashford group	40 people
Bexhill Youth group	24 people
Baslow Road group	60 people

(Q) Advise the youth worker on how many people to survey from each group.

We're on the way

In this section you will:
- find proportional relationships in questions
- use ratios to scale quantities up or down
- calculate missing values in proportional relationships
- state any assumptions you make in order to justify your answer

Take a look: Calories

Declan is on a diet. He counts the calories in his food.

Declan eats beef lasagne at a restaurant.
Declan does not know how many calories are in the lasagne.
He has a ready-made beef lasagne at home.

Here is some of the information on the box.
Individual serving 320 g
Contains 180 calories per 100 g

Declan assumes that the lasagne at the restaurant has $\frac{1}{3}$ more calories per 100 g than the ready meal.
He estimates that the weight of the restaurant lasagne was $1\frac{1}{2}$ times the weight of the ready meal.

Q Use the information to estimate how many calories were in the beef lasagne at the restaurant.

 Here's a possible solution:

$\frac{4}{3} \times 180$ calories $= 240$ calories

❶ '$\frac{1}{3}$ more' means $\frac{4}{3}$ times

A Work out the estimated number of calories per 100 g in the restaurant lasagne

$1.5 \times 320\,g = 480\,g$

A Work out the estimated weight of the restaurant lasagne

$480\,g \div 100 \times 240$ calories $= 1152$ calories

A Work out the estimated number of calories in the restaurant lasagne

The restaurant lasagne contains approximately 1150 calories.

Take a look: Planning a walk

James is planning a walk.

He uses a map to plan a route.
He measures the distance on the map as 42.5 cm.
The scale of his map is 1:50 000
James knows that he walks at an average speed
of 4 miles per hour.

James wants to work out how long his walk will take.

(Q) Work out how long it will take James to
complete his walk.

A There is a mix of units in
the question – cm and miles.
You will need to change the
units so they are the same

 Here's a possible solution:

$42.5 \times 50\,000 = 2\,125\,000$ cm

R Use the ratio to work
out the distance of the walk
in 'real life'. Use the fact that
1 km is $\frac{5}{8}$ of a mile to convert
the units.

$2\,125\,000 \div 100 = 21\,250$ m
$21\,250 \div 1000 = 21.25$ km

A Use the scale on the
map to work out the distance
in real life

$21.25 \times \frac{5}{8} = 13.28125$ miles

I Convert the units from
centimetres to kilometres

$13.28125 \div 4 = 3.320\,312\,5$

A James's walking speed
is given in miles per hour. You
need to convert the distance
into miles

$0.320\,312\,5 \times 60 = 19.218\,75$

A Use the distance and
James's average walking
speed to work out how long
the walk will take

James will take about 3 hours 20 minutes to
complete his walk.

I Change the fraction of
an hour into minutes

I Give the time to a
sensible accuracy

Have a go

4 When Bhavna drives long distances on a motorway, she can drive for 400 miles on one full tank of petrol.
When she drives short distances around town, she can drive for only 200 miles on a full tank of petrol.

It costs £80 to fill Bhavna's car's petrol tank.
This week Bhavna will drive roughly 100 miles around town.
She will also drive roughly 100 miles on the motorway.

Q Estimate how much Bhavna's petrol will cost this week.

5 Mike wants to complete his Duke of Edinburgh Award.
As part of the award he must lead a walking team across part of the North Yorkshire moors.
On a map, part of the walk measures 2.2 cm.

Mike is using a map with a scale of 1:25 000
Mike works out that when he walks 70 strides, he travels a distance of 100 m.
One stride is equal to two normal steps.

Q How many strides must Mike walk to cover 2.2 cm on the map?

6 Mabel, her daughter Jane and Mabel's grandson Tom may have a swine flu injection.
Mabel is 64, Jane is 31 and Tom is 3.

Mabel finds this chart in the local paper.

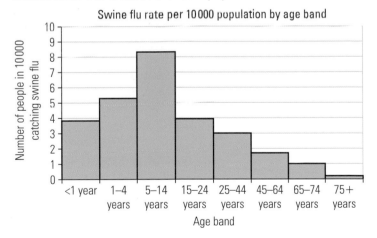

Jane and Mabel decide that someone should have a swine flu injection if the risk of getting swine flu for their age band is more than 1 in 5000.

Q Use the information in the chart to decide who, of Mabel, Jane and Tom, should have a swine flu injection.

Exam ready!

In this section you will:
- find and process information from questions to write ratios
- find and use ratios to solve problems
- find and use proportional relationships to scale quantities up or down
- calculate missing values in proportional relationships
- interpret answers and form conclusions
- state any assumptions made to justify solutions

Take a look: Planning a camping trip

Jackie has to buy the food for the Guides' summer camping trip.
She has this information about last year's trip.

Number of Guides camping last year	Some of the food bought for last year's trip			
	Loaves of bread	Litres of milk	Kilograms of bananas	Tins of beans
43	35	28	42	98
			Total money spent	£214.48

This year 65 Guides are camping.

Q How much money, in total, will Jackie need for the food this year?

💡 Here's a possible solution:

Assume that the Guides this year will eat the same food and the same amounts of food as last year.

The ratio of the number of Guides last year to this year is 43 : 65 •
As a proportion this means that $\frac{65}{43}$ more Guides are camping this year.

$\frac{65}{43} \times £214.48 = £324.21$ •

Jackie will spend roughly £350. •

R The number of Guides is the only information you have to connect the figures for last year and the figures for this year

A First write the information as a ratio

A Use the same proportional value to estimate the increase in money spent on food

I Prices may have increased so Jackie may need to spend more than £324.21

Have a go

7 A company director wants to share out profits of £18 000 as a Christmas bonus.
The director will share out the £18 000 as follows:
   • the nine office workers will each get at least £500
   • the four managers will each get at least twice the bonus of an office worker
   • the director will get at least three times the bonus of an office worker.

Q Work out the bonus amount each person will get.

8 Jim has to buy soft drinks to sell at this year's charity concert.
He has this information about last year's concert:

Number of tickets sold	Number of cans sold	Number of bottled drinks sold
1500	719	430

This year, 2000 concert tickets have been sold.

Q **a)** How many cans and bottles should Jim buy for this year's concert?

A company sells cans and bottles of soft drink in packs of 24.
A pack of cans costs £10.99
A pack of bottles costs £19.92

Jim buys the cans and bottles from this company.
Jim then sells the cans and bottles at the concert.
He expects to sell all the soft drinks he buys.
Jim wants to make a profit of 30% by selling the soft drinks.

Q **b)** Work out how much Jim should charge for each bottle and each can.

You need to decide what to do first. You need to find an estimate for the number of cans and bottled drinks that will be sold at this year's concert.

Now you can:

• Find information in questions and use the information to find ratios and proportions
• Use ratios to divide quantities and use proportions to scale quantities up or down
• Use proportions to compare trends and to predict behaviour

In a Functional Skills Mathematics exam there may be questions where you need to make decisions about which item is the best value.

ResultsPlus
Maximise your marks

Pat has a dog called Athena. Athena weighs 35 kg.
Pat collects information on two types of tinned dog food.

Dog Food 1	Dog Food 2
Recommended daily use according to dog weight: 30 kg $1\frac{1}{4}$ tins 40 kg $1\frac{1}{2}$ tins	**Recommended daily use according to dog weight:** 30 kg $1\frac{1}{2}$ tins 40 kg 2 tins
Tin weight: 1.2 kg **Price:** £1.25	**Tin weight:** 825 g **Price:** 90p

Which dog food is the best value for Pat to buy?

Exam tips

Although the weight of each tin is given, this information is not needed to work out the best value.
You will have to decide how many tins of food Athena should eat each day.

Student response	Exam comments
Let's look at a poor answer: Dog food 1 because both types need two tins bought to feed Athena. After one feeding of dog food 1 there will be enough food left in the second tin to feed Athena again.	The student has not found out how much money it will cost to feed Athena each day. The student has made a decision about the quantity of dog food needed, but does not seem to have thought about Athena's weight
Let's look at an answer that will gain some marks: **Dog Food 1** **Dog Food 2** 30 kg = $1\frac{1}{4}$ tins 35 kg = $1\frac{3}{4}$ tins 40 kg = $1\frac{1}{2}$ tins 35 kg = $1\frac{1}{3}$ tins Dog food 2 is the best value for Pat as it is cheaper. 40 ÷ 2 = 20 × 1 = 20 30 ÷ 2 = 15 × 1 = 15 17.5 is the middle number.	This student has thought about how much dog food Athena needs each day. The reference to $1\frac{1}{3}$ tins shows that the student is considering the amount of dog food Athena should have. However, the calculations do not compare the costs of food 1 and food 2.
Let's look at a complete answer: **Dog Food 1** (1.25 + 1.5) ÷ 2 = 1.375 tins 1.375 × 1.25 = £1.72 per day **Dog Food 2** (1.5 + 2) ÷ 2 = 1.75 tins 1.75 × 0.90 = £1.575 per day Dog Food 2 is cheaper each day.	This student has found the correct amount of food that Athena needs each day by converting the fractions to decimals and finding the average number of tins needed. The costs are correct for both tins. The student has made a decision based upon comparing the two costs.

Exam question

Barbara wants to find out the number of tins of milk powder she needs to feed her two babies for one week.

She also wants to know the cost of the milk powder.

The milk is made by mixing the milk powder with water.

These are the instructions on a tin of milk powder:

> Recommended amount of milk needed per day
>
> 2.6 fluid ounces per pound of body weight
>
> Use one 4.5 g scoop of powder to make 1 fluid ounce of milk.
>
> 1 fluid ounce = 30 ml

There are **900 g** of milk powder in each tin.

Each tin of milk powder costs £7.97

The total weight of Barbara's two babies is **27 pounds**.

How many tins of milk powder does Barbara need for one week?
How much in total will the tins cost? Show clearly how you get your answer.   (8 marks, May 2009)

How students answered

■ 40% of students (0–2 marks)	● 30% of students (3–5 marks)	▲ 30% of students (6–8 marks)
Some students did **not** use multiplication and division correctly. They did **not** understand which operation to use when calculating the number of scoops needed, based on body weight. Some students did not identify '1 fluid ounce = 30 ml' as redundant information. Poor presentation of working was also a problem.	Some students made progress. However, these students: • did **not** check their answers to see if they made sense • had difficulty converting between different unit systems • rounded incorrectly.	Some students were able to round correctly and/or state the correct cost. Some students did **not** show enough evidence of following a method.

Put it into practice

1 Compare the costs of the same food items sold by different supermarkets.

2 Look at liquid or powder foods for people or for plants and calculate the costs of buying these foods over a period of time.

ASSESSMENT PRACTICE 1

1 A company needs to hire some new machines.

They will choose one of four types of machine.

Machine type	Hire cost per week	Items made per week
A	£1600	400
B	£7430	3600
C	£13 800	6200
D	£3720	1600

The company have an order for 10 000 items.
They have 7 weeks to make the items.

(Q) Advise the company on which machine to hire.

2 Jess records the amount of water used by her family.

She estimates that they use 100 litres of water for drinking and cooking per week. Other water used by her family is shown in the table below.

Activity	Average number of uses per week	Litres used each time
bath	17	80 per bath
flushing the toilet	103	8 per flush
shower	17	60 per shower
washing machine	15	65 per wash
dishwasher	10	25 per wash

(Q) a) How much water does Jess's family use in one week?

Jess wants everyone in her household of four people to use less water. She would like them each to use, on average, no more then 100 litres of water per day.

(Q) b) How much less water on average each week will the family have to use?

3 Henry feeds his dog Complete Dog Food. His dog weighs 35 kg.

The normal price of the dog food is given here.

A supermarket has a special offer on packs of 5 kg Complete Dog Food.

Henry wants to buy enough Complete Dog Food to feed his dog for 6 months.
He usually buys the dog food in 3 kg packs.

> Offer valid from 22 January for 30 days
>
> *Buy one 5 kg pack of*
> ## Complete Dog Food
> *and get the second half price*

COMPLETE DOG FOOD

Pack weight	Price
1 kg	£1.47
3 kg	£3.08
5 kg	£4.30

Recommended daily use for dog weight of 35 kg: 500 g per day

(Q) How much money will Henry save if he buys enough Complete Dog Food for 6 months in 5 kg packs instead of 3 kg packs?

4   Sally has two children, aged 8 and 13. Sally and the children are going on
    holiday to Tenerife. Sally wants to take her children to some parks in Tenerife.

    The costs of visiting two parks are shown in the table below.

Ticket type	Cost (Euros) Children aged 3–11 years	Cost (Euros) Adults and children 12 years or older
Lora Parque	20.50	30
Siam Park	18	28
Twin ticket	33	55

    A twin ticket means that people can go to both Lora Parque and Siam Park.
    If Sally buys the tickets online before she goes on holiday she will save
    5 euros on each ticket she buys.

    Sally has £300 to spend. She wants to take her children to both parks.
    The exchange rate is £1 = €1.10

Q a) Which tickets should Sally buy?
       b) How much does Sally save by buying these tickets rather than buying separate
           tickets when she arrives at each park?

5   A cinema starts a reward scheme.
    A customer will get 10 points for every full pound they spend at the cinema on each visit.
    Simon visits the cinema four times each calendar month. He watches two films on each
    visit. He buys a ticket that costs £7.50 for each film. He also buys a small soft drink and a
    small popcorn each time he visits the cinema.

    The total cost of the soft drink and the popcorn is £4.30

Q a) How many points does Simon get in one month?

       Customers can use their points to pay for film tickets, soft drinks and popcorn.
       Standard seat ticket costs 800 points
       Small soft drink costs 300 points
       Small popcorn costs 400 points

       Simon collects points for one year.
       He then decides to use all his points.

       He still visits the cinema four times a month and buys two standard seat tickets, a small
       drink and a small popcorn each time.
       Simon wants to know how long his points will last.

Q b) Work out how many cinema visits Simon will be able to pay for with his points.

6   Bob needs to calculate his National Insurance contributions and income tax bill for the
    year. He estimates that he will have made £22 000 profit.

    National Insurance Contributions
    A flat rate of £2.40 per week plus 8% in the pound for every £1 of profit over £5715

    Income Tax
    20% in the pound for every £1 of profit over £6475

Q Work out the total amount of money Bob will pay in National Insurance
       contributions and income tax.

4 Time
Know Zone

You cannot get away from time. It runs through our lives. Getting to an appointment on time, planning your holiday, catching a bus to the cinema – we work with time every day.

Key facts about time

You need to know some key facts about **time**.

60 seconds	=	1 minute	52 weeks	=	1 year
60 minutes	=	1 hour	12 months	=	1 year
24 hours	=	1 day	10 years	=	1 decade
7 days	=	1 week	100 years	=	1 century

You can write a time in different ways. You can use either the 12 hour clock or the 24 hour clock.

12 hour clock		24 hour clock	
Correct	Incorrect	Correct	Incorrect
08:30 am	830	17:20	720
8.30 am	8.3 am	17.20	7.2
		1720	1720 pm

A 'time interval' is the length of time between two different times.
It can be written in different ways.

Correct	Incorrect
3 hours 20 minutes	3.2
3 hr 20 mins	3.2 hours
3 20	3 h 2
200 minutes	3.2 min

It is useful to learn the number of days in each month.

Steve got to work 25 minutes before Jess.
Jess got to work at 0910.
What time did Steve arrive?

> Count back 10 minutes from 0910 to 0900
> Count back 15 minutes from 0900 to 0845
> Remember there are 60 minutes in 1 hour

Steve got to work at 0845.

Let's get started

In this section you will:
- ◉ learn to recognise information you do not need
- ◉ add together length of time
- ◉ learn to write the time properly

🔍 Take a look: Journey time

Ursa plans to travel from York to Edinburgh.
She wants to find out how long the journey will take.
She looks at her train timetable:

King's Cross	13:30
York	15:34
Durham	16:25
Edinburgh	18:25

R The information about King's Cross and Durham is **not** needed

Q How long will Ursa's journey take?

💡 Here's a possible solution:

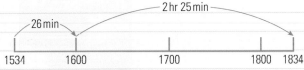

Ursa's journey time is:
26 min + 2 hr 25 min = 2 hr 51 min

A You can 'add on' from 1534: 1534 to 1600 is 26 min 1600 to 1825 is 2 hr 25 min 26 min + 2 hr 25 min = 2 hr 51 min

◎ Have a go

1 A plane should leave Tenerife Airport at 2040.
It should arrive at Manchester Airport at 0050 the next day.
The plane leaves Tenerife Airport late. Now the plane will arrive at Manchester Airport 2 hours 40 minutes late.

Q At what time should the plane now arrive at Manchester Airport?

2 Mr Potts asks a builder to build a garage for him.
The garage must be finished on or before Monday 4 June 2012.
The builder works out that he needs 21 working days.
He works from Monday to Thursday inclusive.

Q What is the latest date that the builder can start to build the garage?

We're on the way

In this section you will:
- ◉ make a plan of a schedule for a situation
- ◉ use units of time
- ◉ plan different events that happen at the same time
- ◉ design a programme

 Take a look: Festival

John is going to play at a music festival.
John must play the piano for between 7 and 8 minutes.
He wants to play one Invention, one Sinfonia and one Partita.

R You need to suggest one piece of music from each column. All the times added together must be between 7 and 8 minutes

John can play the following pieces of music by J S Bach.

Inventions		Sinfonias		Partitas	
No. 6	4:04	No. 2	2:11	No. 2	2:51
No. 9	3:31	No. 3	1:27	No. 6	2:59
No. 10	0:58	No. 10	1:08	No. 10	0:58
		No. 12	1:26		

Key: 3:31 = 3 minutes 31 seconds

Q Suggest three of these pieces of music that John could play. Show all of your working.

Q Here's a possible solution:

$$3:31 + 1:27 + 2:51 = 6 \text{ min } 109 \text{ sec}$$
$$= 7 \text{ min } 49 \text{ sec}$$

A Change 109 seconds into 1 minute 49 seconds

John could play Invention No. 9, Sinfonia No. 3 and Partita No. 2.

Take a look: Cinema

A multi-screen cinema is showing some films on Saturday.

The cinema will show one of the films at these start times:

11.50	12.30	13.45	14.30
15.15	16.15	17.00	18.00

The film is 130 minutes long.
The cinema must show the film on the smallest possible number of screens.

Q Make a plan for the showings of this film on Saturday.

Here's a possible solution:

130 minutes = 2 hours 10 minutes

Start	Finish	Start	Finish
11.50	14.00	15.15	17.25
12.30	14.40	16.15	18.25
13.45	15.55	17.00	19.10
14.30	16.40	18.00	20.10

A Work out the finishing time for each showing of the film. This will help you to find out how many screens are needed. The first showing ends at 14.00, so put the 12.30 and 13.45 showings on other screens. The 14.30 showing will be shown on the same screen as the 11.50 showing

SCREEN PLAN
Screen 1
11.50 to 14.00   14.30 to 16.40   17.00 to 19.10
Screen 2
12.30 to 14.40   15.15 to 17.25   18.00 to 20.10
Screen 3
13.45 to 15.55   16.15 to 18.25

R Think about the smallest number of screens to run the film

I Make sure that people will understand the screen plan

Have a go

3 Adam is a carpet layer. He can normally lay 25 m² of carpet per day in a house. He estimates that he will be able to work 50% quicker when he lays carpets in an office.

Adam gets a job laying 175 m² of carpet in an office.

Q How many days will it take Adam to finish laying the carpet?

4 Supermarkets are very busy on 23 December.
The manager of a supermarket asks all her staff if they will work an extra 4-hour shift on 23 December.
The manager draws this table.

Time	Number of extra staff needed
10.00–11.30	4
11.30–14.00	6
14.00–16.30	4
16.30–18.00	6

Q Plan a rota of 4-hour shifts so that the supermarket has enough staff.

Exam ready!

In this section you will:

- ⦿ use suitable forms to identify the problem
- ⦿ change and improve the features of a programme
- ⦿ show your programme clearly

Take a look: Competition

There are 14 competitors in stage 1 of a competition.
Each competitor has to perform for 25 minutes.

The competition venue is open from 10.00 am for 12 hours each day.
The venue is booked for one day.

The judges will need

- a 100 minute break for judging and lunch
- a 100 minute break for judging and dinner
- three 20 minute breaks.

ResultsPlus
Exam Tip

Features of a good programme

- Programme times do not overlap.
- Time for each competitor's performance is managed.
- Break times are sensibly spread out through the day.
- All competitors can compete within the time period allowed.
- Competition is between 10.00 am and 10.00 pm.
- The finished programme is easy to understand.

Q Design an events programme for stage 1 of the competition.
Your programme should show the start and end time for each competitor

Here's a possible solution:

Venue open for 12 hr
14 competitors

Each competitor needs 25 min (or more)
Try 30 min each × 14 = 7 hr

Judging and lunch break = 100 min (1 hr 40 min)

Judging and dinner break = 100 min (1 hr 40 min)

3 breaks = 3 × 20 min = 1 hr

Total time = 11 hr 20 min

Break

C8 15.30–16.00
C7 15.00–15.30
C6 14.30–15.00

C9 16.20–16.50
C10 16.50–17.20

Lunch

Dinner

C5 12.20–12.50
C4 11.50–12.20

Events programme

C11 19.00–19.30
C12 19.30–20.00

Break

Break

C3 11.00–11.30
C2 10.30–11.00
C1 10.00–10.30

C13 20.20–20.50
C14 20.50–21.20

Start

End

Have a go

5 This railway timetable is from Monday to Friday.

York		0836		0852		0935		1006
Leeds		0908		0949		1008		1053
Skipton	0708		0747		0756		0843	
Leeds	0805		0840		0905		0940	
Wakefield	0817		0852		0918		0952	
Doncaster	0836		0911	0916	0935	1003	1014	1030
Doncaster	0837		0912	0917	0936	1004	1015	1030
Retford					0951			

Samil needs to travel from York to Doncaster.

(Q) Plan a journey for Samil.
Give the times the train or trains you choose leave and arrive at the station.

6 Cara is a journalist.
Her employer should:
- pay her claims for expenses within 30 days
- pay her claims for work within 40 days.

Cara records information about the payments her employer puts into her bank account.

Date of claim	Claim	Amount	Date of payment
10 April	E	£24.70	15 May
1 April	W	£1250.00	8 May
23 March	W	£87.50	8 May
18 March	E	£138.00	11 April
5 March	W	£550.00	4 April
4 March	E	£67.50	4 April
2 March	E	£320.00	27 March
2 March	W	£1200.00	27 March

Key: E = Expenses   W = Work

(Q) Analyse how well Cara's employer manages the payments for her expenses and work.

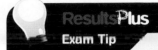

ResultsPlus Exam Tip

When you are asked to describe something it means you should explain whatever you see as fully as you can without repeating yourself. Your answer needs to be interpreted by someone else. It must be unambiguous.

Now you can:

- Write time in the 12 and 24 hour clock
- Write periods of time
- Calculate with time
- Find information from timetables and schedules
- Find solutions to problems when time is a key part of the problems

Know Zone

We use different types of measurement in everyday life. Temperatures are given in weather forecasts. We measure our own weight and height. You need to know which **units** are used for different measurements and how to work with each unit.

In this chapter you will work with measures for:

- temperature
- length
- weight
- capacity

Temperature, Length, Weight, Capacity

Temperature is usually recorded in degrees Celsius (°C), but sometimes you may see degrees Fahrenheit (°F) used. The approximate melting point of ice is 0°C. The approximate boiling point of water is 100°C.

Length is a measure of **distance**. The most common units for measuring distance are:

Metric	Imperial
millimetres (mm)	inches (in)
centimetres (cm)	feet (ft)
metres (m)	yards (yd)
kilometres (km)	**miles** (mile)

Weight is a measure of how heavy an object is. The most common units used to measure weight are:

Metric	Imperial
milligrams (mg)	ounces (oz)
grams (g)	**pounds** (lb)
kilograms (kg)	stones (st)
tonnes (t)	tons (t)

Capacity is a measure of **volume**. The volume of a container is the amount of space in it. The capacity of a container is the amount of fluid it can hold. The most common units for measuring capacity are:

Metric	Imperial
millilitres (ml)	fluid ounces (fl oz)
centilitres (cl)	**pints** (pt)
litres (l)	**gallons** (gal)

Metric and Imperial units

Measurements are most often given in **metric** units. The metric system is based on tens, hundreds and thousands. However, **imperial** units are still used in the UK. For example, distances on road signs are given in miles and drinks can be measured in pints. You need to know the connections between the metric units for length, weight and capacity. You need to be able to work with the equivalent imperial units.

	Length	Weight	Capacity
Metric	10 mm = 1 cm 100 cm = 1 m 1000 m = 1 km	1000 mg = 1 g 1000 g = 1 kg 1000 kg = 1 tonne	10 ml = 1 cl 100 cl = 1 litre 1000 cm^3 = 1 litre
Imperial	12 in = 1 ft 3 ft = 1 yd 1760 yd = 1 mile	16 oz = 1 lb 14 lb = 1 stone	20 fl oz = 1 pint 8 pints = 1 gallon

A **conversion factor** is a number you multiply or divide by to change from one unit to a different unit. In exam questions using conversion between metric and imperial you may be asked to use specific conversion factors. The table below shows some conversion factors that are often used.

Length			Weight			Capacity		
metric		imperial	metric		imperial	metric		imperial
2.5 cm	=	1 in	25 g	=	1 oz	1 litre	=	$1\frac{3}{4}$ pint
30 cm	=	1 ft	1 kg	=	2.2 lb	4.5 litre	=	1 gallon
1 m	=	39 in						
8 km	=	5 miles						

Compound measures

A **compound** measure is a measure that relies on two other measures.
You could be given the **formula** for a compound measure to work with in an exam question.
Speed is an example of a compound measure..

Speed is a measure of the distance travelled in unit time. The units commonly used for speed are metres per second (m/s), kilometres per hour (km/h) and miles per hour (mph).

You can use this formula to see the connections between distance, speed and time.

$$\text{speed} = \frac{\text{distance}}{\text{time}} \qquad \text{distance} = \text{speed} \times \text{time} \qquad \text{time} = \frac{\text{distance}}{\text{speed}}$$

Use the tables in the Know Zone to answer the questions in this chapter.

Let's get started

In this section you will:
- learn to read and use scales
- use scales to estimate, measure and compare length, distance, weight, capacity and temperature

Take a look: Change in temperature

The thermometer shows the **maximum** and **minimum** temperatures, in degrees Celsius, recorded in Moscow one day.

(Q) What is the difference between the maximum temperature and the minimum temperature?

(💡) Here's a possible solution:

On this thermometer, ten marks on the scale represent 5°C, so one mark represents $5 \div 10 = 0.5°C$.

The scale shows that the maximum temperature is 2.5°C and the minimum temperature is −4.5°C.

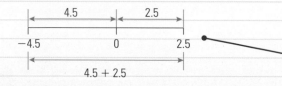

The difference between the two temperatures is $4.5 + 2.5 = 7°C$.

(A) First work out what each mark on the scale represents

(I) Sketch a number line

Have a go

1   This is a metre scale.
    The arrows show the lengths of two different pieces of wood.

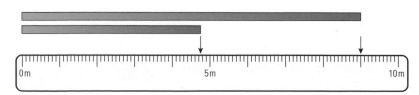

(Q) What is the difference in length of the two pieces of wood?

2 A car's fuel tank holds 40 litres. The fuel gauges on the right show the petrol in the tank at the start of the journey and at the end of the journey.

Start   End

 Approximately how much petrol did the car use during the journey?

3 The thermometer shows the temperature inside a freezer before it is defrosted and the temperature after it has been defrosted.

 What is the difference between the two temperatures?

○ We're on the way

In this section you will:
- ◉ use units of measure within the same system
- ◉ calculate with units of measure across systems
- ◉ learn to identify the correct conversion factors to work with

🔍 Take a look: Motorway safety barriers

The length of each section of a safety barrier on a UK motorway is 30 metres. The sections are placed end to end to form the barrier. There are two barriers in the centre of a UK motorway.

 How many sections are needed for 10 km of motorway?

💡 Here's a possible solution:

$10\,km = 10 \times 1000 = 10\,000\,m$

> **A** Make sure all of the values are in the same units

The number of sections needed is $10\,000 \div 30 = 333.33\ldots$

This means that 333 sections will not be enough, so 334 are needed. There are two barriers, so the total number of sections needed is 668.

> **❶** In practical questions you often need to round your answers up or down to a whole number

Have a go

4 Mark is making drinks to sell at a college sports day.
He mixes 1 litre of orange squash with water to make between 10 and 12 litres of drink.
He sells the drinks in 200 ml glasses.

Q Estimate the number of glasses of drink Mark can make from 1 litre of orange squash.

5 A case of 24 cans of cola weighs 7.92 kg.
There is enough space in a lorry for 8000 cases of cola.
However, the lorry can only carry a maximum weight of 60 tonnes.

Q What is the maximum number of cans of cola the lorry can carry?

6 A town council puts solar lights along the centre of a dual carriageway. The length of the dual carriageway is 11 km. The distance between each of the solar lights is 5.5 m. The solar lights cost £1610 each.

Think First!

Draw a diagram to help you see the relationship between the distance and the number of solar lights.

Q How much does it cost the council to put the solar lights on the dual carriageway?

Take a look: Height conversion

Jessica is 5 foot 4 inches tall.

Q What is her height in metres?

Q Here's a possible solution:

1 foot = 30 cm so 5 feet = 5 × 30 = 150 cm
1 inch is 2.5 cm so 4 inches = 4 × 2.5 = 10 cm

R Identify the conversion factors you need

Total height = 150 + 10 = 160 cm = 1.6 metres

Have a go

7 Franck wants to raise money for charity.
He is going to make some cupcakes to sell.

A recipe uses 7 oz of flour. The recipe makes
between 12 and 16 cupcakes. Franck buys 1 kg of flour.

(Q) **Estimate the number of cupcakes he can make.**

> **Think First!**
>
> First identify the
> conversion factors you
> need. Make sure you are
> working with the correct
> units in these questions.

8 Paul and Patrick are going to cook a meal for the family.
They buy a large turkey weighing 25 pounds.
Their recipe book says:
- for a turkey under 4.5 kg, allow 45 minutes cooking time per kg plus 20 minutes
- for a turkey between 4.5 kg and 6.5 kg, allow 40 minutes cooking time per kg
- for a turkey over 6.5 kg, allow 35 minutes cooking time per kg.

(Q) **How long will the turkey take to cook?**

9 A farmer expects each of his cows to produce between
7000 and 8000 litres of milk per year.
He sells milk to supermarkets in 6-pint bottles.

(Q) **Estimate the number of 6-pint bottles the farmer can fill
with milk from one cow in a year.**

> **Think First!**
>
> Sometimes it is easier to
> convert to one unit rather
> than the other when
> tackling a question.

10 Hamid wants to fly from London to New York. He finds out that it is 5600 km from
London to New York. He has a token for a free flight for a distance of up to 3000 miles.

(Q) **Can Hamid use his token for a flight from London to New York?**

Exam ready!

In this section you will:
- learn to identify conversion factors
- learn to work with compound measures

Take a look: Working with speed

A car travels at a speed of 60 mph.

(Q) **What is this speed in km per hour?**

(💡) Here's a possible solution:

5 miles = 8 kilometres
1 mile = $\frac{8}{5}$ kilometres = 1.6 kilometres
The conversion factor from miles to kilometres is 1.6 ●————

> **A** Work out the necessary
> conversion factor

60 mph = 60 × 1.6 = 96 km per hour

Have a go

11 Khalid is going to drive himself and two friends to the Glastonbury festival and home again. He will drive 217 miles to get to Glastonbury.

Khalid's car travels between 46 and 63 miles to the gallon. Petrol costs 102.9 p per litre. Khalid and his friends will share the petrol costs.

Q Estimate how much it will cost each person for the petrol to Glastonbury and home again.

12 The table shows the world population in **millions** from 1960 to 2010.

Year	1960	1970	1980	1990	2000	2010
Population (millions)	2982	3692	4435	5263	6070	6803

Q **a)** Draw a suitable graph or chart to display this information.

b) Estimate the population of the world in 2020.

The formula to calculate population density is:

$$\text{Population density} = \frac{\text{total number of people}}{\text{total land area}}$$

The total population of the UK in 2010 was approximately 60.6 million.
The total land area of the UK is 245 000 km².
The total land area of the world is 150 million km².

c) What is the difference between the population density of the world and the population density of the UK in 2010?

13 Emma has a water meter in her flat.
Every 5 litres of water used costs 1p.

Emma investigates the amount of water she uses:

Activity	Amount of water used	In an average week
Toilet	9 litres per flush	5 flushes per day
Washing machine	80 litres each time	once a week
Dishwasher	25 litres each time	twice a week
Washing up	10 litres each time	once a day
Bath	60 litres each bath	twice a day

Q How much money could Emma save on her water bill if she:
a) changes her toilet to one which uses 4–5 litres per flush
b) has a shower twice a day instead of a bath (a standard shower uses 9 litres of water per minute)?

14 Jose makes deliveries to Birmingham to Coventry, Nottingham, Sheffield and Stoke-on-Trent.

The table shows the distances between each of these towns.

Birmingham

22	Coventry			
52	56	Nottingham		
90	95	44	Sheffield	
48	63	55	50	Stoke-on-Trent

Key: 60 = 60 miles

Jose starts each day in Birmingham. He makes deliveries to each of the towns. Then he drives back to Birmingham.
Jose drives at an average speed of 45 mph.

Think First!

mph means miles per hour $= \dfrac{\text{miles}}{\text{hours}}$

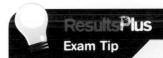

ResultsPlus
Exam Tip

It is important that you set out your working clearly. Any plans must be easy for someone else to follow.

Jose takes between 10 and 15 minutes to make deliveries at each town.
He has 1 hour for his lunch.

(Q) Plan the shortest route for Jose to take and work out the total time he will be away from Birmingham.

Now you can:

- Read and use scales
- Use scales to estimate, measure and compare length, distance, weight, capacity and temperature
- Use mixed units of measure within the same system
- Calculate with units of measure between systems
- Identify the correct conversion factors to work with
- Work with compound measures

6 2D representations of 3D objects
Know Zone

There are many two-dimensional (2D) representations of three-dimensional (3D) shapes in everyday life. For example, a kitchen planner uses a 2D plan of the kitchen floor to show how the 3D kitchen units will fit into the room. When you are doing tasks using 3D objects it helps if you represent the problem in 2D.

3D shapes

You will need to know the following 3D shapes. You also need to know the number of faces, edges and vertices (corners) they each have.

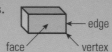

Fact

Vertices is the plural of vertex.

3D shape	Number of faces	Number of edges	Number of vertices
cube	6	12	8
cuboid	6	12	8
cylinder	3	2	0
square-based pyramid	5	8	5
triangular prism	5	9	6

A net is a 2D shape that you can fold into a 3D solid. Most solids have more than one possible net. Some examples of nets for common 3D shapes are given below.

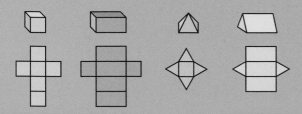

You can use isometric paper to represent 3D objects in 2D.

This isometric paper shows a representation of a cuboid that is 3 cm by 4 cm by 5 cm (not to scale).

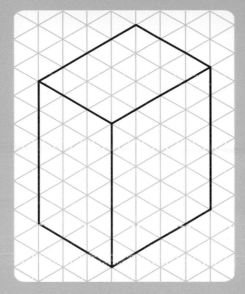

When architects design buildings, they produce 2D drawings to show what the building will look like from each side. These drawings are called plans and elevations. The view from the top is the plan. The views from the front and sides are the elevations.

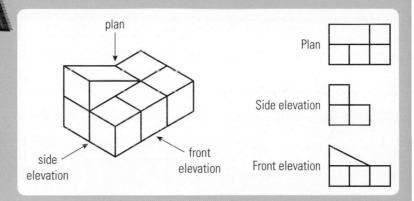

plan

side elevation

front elevation

Plan

Side elevation

Front elevation

Plans and elevations are also very useful to use when planning the design of a room.

These diagrams show the plan and one of the side elevations of a kitchen.

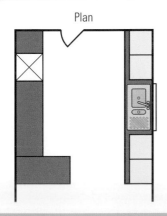

Plan

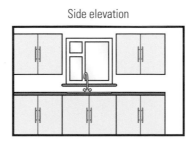

Side elevation

Let's get started

In this section you will:
- learn to recognise and use common 2D representations of 3D objects

🔍 Take a look: Packaging problem

A tin of shoe polish is 8 cm in diameter and 2 cm high. The shoe polish must be stored in an upright position. The tins of shoe polish are packed into a carton that is 40 cm long by 32 cm wide by 10 cm high.

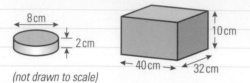

(not drawn to scale)

Q How many tins of shoe polish will fit into the carton?

💡 Here's a possible solution:

The base of the carton is 40 cm by 32 cm.
The tins are 8 cm wide.

> **A** Examine the relationship between the lengths of the shoe polish tins and the lengths of the carton

This means $40 \div 8 = 5$ tins will fit along one edge and $32 \div 8 = 4$ tins will fit along the other edge. A total of $5 \times 4 = 20$ tins will fit on the bottom of the carton.

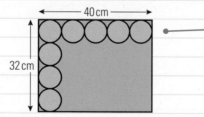

> **R** Make a sketch of the 3D problem in 2D

The carton is 10 cm high. The tins are 2 cm high.

This means $10 \div 2 = 5$ tins will fit on top of each other in the carton.

The total number of tins of shoe polish that will fit in the carton is $20 \times 5 = 100$.

> **I** Use the relationship to work out the total number of tins that will fit in the carton

 **Have a go**

1   A company makes circular drink mats. Each mat has a radius of
    4.5 cm and is 3 mm thick. The company wants to pack the mats
    into rectangular boxes. They will pack 12 mats into each box.

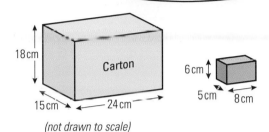

(Q) **Work out the minimum dimensions of the box.**

2   A carton is 24 cm by 15 cm by 18 cm.
    The carton will be packed with boxes of staples.
    The boxes are 8 cm by 5 cm by 6 cm.

(Q) **What is the largest number of boxes of staples
    that will fit in the carton?**

(not drawn to scale)

3   A company makes butter and needs a drawing of a packet
    of butter for a poster.
    Each packet of butter is a cuboid. It has dimensions 8 cm by 6 cm by 3 cm.

(Q) **Draw this cuboid on a centimetre isometric grid.**

4   Gemma is making a six-sided dice for a game.
    She has drawn a net and filled in some of the numbers.
    On a standard dice the numbers on opposite sides
    add up to seven.

Fact

There are 11
different possible
nets for a cube.

(Q) **Copy the net and fill in the missing numbers on the remaining sides.**

○ We're on the way

In this section you will:
  ◉   learn to solve problems involving 2D shapes and parallel lines

🔍 Take a look: Paving a patio

The diagram shows the plan of Peter's patio.
Peter is going to pave the patio with square
paving slabs.
Each paving slab has sides of 0.5 m.

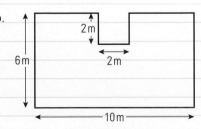

(not drawn to scale)

(Q) **How many paving slabs does Peter need?**

Here's a possible solution:

For the 10 m width, **20** slabs are needed. •

For the 6 m width, **12** slabs are needed. •

The number of slabs needed for the whole
rectangle is 20 × 12 = **240**. •

The cut-out piece is 2 m by 2 m.
This would need 4 × 4 = **16** slabs. •

The total number of slabs needed is
240 − 16 = **224** paving slabs. •

Ⓐ Examine the relationship
between the lengths of the
paving slabs and the lengths of
the patio – the slabs are 0.5 m
long, so two slabs are needed
for every metre of patio

Ⓡ Work out how many
paving slabs are needed for
the full rectangle. Then take
away the number of slabs
that would be needed for the
cut-out piece of patio

Ⓘ Use the relationship to
work out the total number of
paving slabs Peter needs for
the patio

Have a go

5 The diagram shows a worktop in a kitchen.
Asif is going to cover the worktop with square tiles. Each tile is 4 cm by
4 cm.

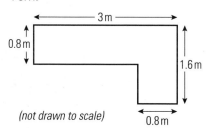

3 m

0.8 m

1.6 m

0.8 m

(not drawn to scale)

Think First!

First convert all of the
lengths to the same units.

Ⓠ How many tiles does Asif need to cover the worktop?

6 The diagram shows the floor plan of
a bedroom. Naznin is going to cover the
floor with carpet. The width of the carpet
is 4 m.

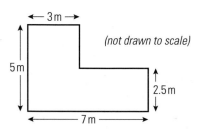

3 m

(not drawn to scale)

5 m

2.5 m

7 m

Ⓠ **a)** Draw a scale diagram to show how the carpet could be laid.
b) Naznin wants to cover the floor completely with carpet.
Work out the shortest length of carpet she needs.

7 The diagram shows one side elevation of a dining room. Jacky is going to cover the wall with wallpaper. One roll of wallpaper is 52 cm wide and 10 m long.

When you put wallpaper on a wall you need to allow an extra 50 mm of wallpaper at the top and bottom of the wall. The sides of the pieces of wallpaper should touch each other, but they should not overlap.

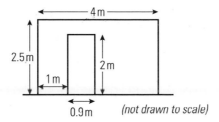

(not drawn to scale)

Exam Tip

You need to decide what the scale will be.
You need to clearly communicate what the scale is.

Q **a)** Draw a scale diagram to show how Jacky could hang the wallpaper on the wall.

b) How many rolls of wallpaper does Jacky need?

c) What length of wallpaper is left over?

Exam ready!

In this section you will:
- learn to apply 2D representations of 3D shapes in more practical situations

Take a look: Moving house

Alphonce is moving out of his student flat for the summer holidays.
He packs his possessions into packing boxes.
The packing boxes are 51 cm long, by 41 cm wide, by 61 cm high.

Alphonce rents a van to move the packing boxes.
The dimensions of the interior of the van are:
Length 2.4 m
Width 1.7 m
Height 1.4 m

Alphonce has to keep the boxes upright.
He can stack the boxes on top of each other.

Q **a)** What is the maximum number of packing boxes that Alphonce can fit into the van?

The van costs £60 to rent. Boxes cost 50p each.

b) Alphonce uses the maximum number of packing boxes.
How much does his move cost altogether?

Here's a possible solution:

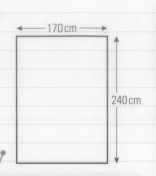

a) Number of boxes that will fit along the length of the van:

$240 \div 51 = 4.7$ so **4** boxes will fit along the length

Number of boxes that will fit along the width of the van:

$170 \div 41 = 4.1$ so **4** boxes will fit along the width.

R Convert lengths to the same units and make a sketch of the floor plan of the van

The height of the van is 140 cm, $140 \div 61 = 2.2$ so **2** boxes can be stacked on top of each other.

The total number of boxes that will fit in the van is $4 \times 4 \times 2 = $ **32**

A Examine the relationship between the lengths of the packing boxes and the dimensions of the van to work out how many boxes will fit along the length, the width and the height of the van

b) Total cost (£) $= 60 + 0.5 \times$ number of boxes
$= 60 + 0.5 \times 32$
$= 60 + 16$
$= 76$
The move costs Alphonce £76

I Use your answers to work out the total number of boxes

A Convert 50p to pounds

 Have a go

8 The diagram shows a plan of Aseem's garden.
Aseem plans a new garden design. He wants to have:
- at least 20 m² of lawn turf
- at least 5 m² of paving stones.
The rest of the garden will be flower beds.

The cost of lawn turf is £2.24 per m². The lawn turf comes in rolls that are 40 cm wide by 250 cm wide. Paving stones measure 0.5 m by 0.5 m and cost £8.75 each.
Aseem has a budget of £400.

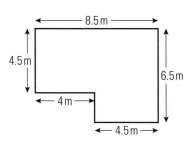

Think First!

When you draw your scale diagram, choose a scale that is easy to work with.

 a) Draw a scale diagram to show a possible design for Aseem's garden.
b) Calculate the total cost of your garden design.

9 Mia wants to fit new units in her kitchen. All of the units have a depth (distance from front to back) of 60 cm.
The table shows the width and the price of each unit.

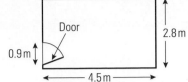

Width	50 cm	60 cm	80 cm	100 cm
Price	£53.00	£55.50	£59.50	£68.00

Mia wants to fit as many units as possible along the walls of her kitchen. She has an oven of width 60 cm and a fridge of width 60 cm. Both the oven and fridge have a depth of 60 cm. Her sink needs to fit into a 60 cm unit.
Mia will keep the oven, the fridge and the sink in the kitchen.
She has a budget of £600.

Think First!

Remember to put in the kitchen units so that the doors can open.

Q **a)** Draw a scale diagram to show how Mia could fit the units in her kitchen.

b) Calculate the total cost of these units.

10 The diagram below shows a side elevation of Kathryn's lounge.
Kathryn wants to paint the wall.

Q **a)** What is the total area of wall Kathryn needs to paint?

The wall needs two coats of paint.
Kathryn can buy the paint in 5-litre tins or 2.5-litre tins or 1-litre tins.
She can paint up to 14 m² with 1 litre of paint.

Think First!

Remember that the window and door will **not** be painted and Kathryn will paint the wall twice.

Q **b)** Which size tin of paint should Kathryn buy to paint the wall? You must show all your working.

11 A playgroup uses large cubes for children to sit on.
The cubes have letters on each face.
One of these cubes is shown on the right.

The bottom face of the cube has the letter T written on it.
The other two hidden faces of the cube have O and Z written on them.

Q Draw a possible net of the cube. Put the letters on the net.

Now you can:

- Recognise and use common 2D representations of 3D objects
- Solve problems involving 2D shapes and parallel lines
- Apply 2D representations of 3D shapes in more practical situations

7 Formulae and equations
Know Zone

We can use formulae to describe the relationship between quantities. Engineers and scientists use formulae to invent new technology and explain how the world works. People also use formulae in their everyday lives to solve many day-to-day problems, from calculating wages to cooking roast dinners.

We can write formulae using words or **algebra**. For example, the **formula** for speed is:

$$\text{speed} = \frac{\text{distance}}{\text{time}} \quad \text{or} \quad S = \frac{d}{t} \quad \text{where} \quad S = \text{speed} \quad d = \text{distance travelled}$$
$$t = \text{time taken}$$

In many jobs and everyday situations people need to use formulae. It is important to know how to recognise and apply formulae from unfamiliar contexts.

When you use formulae you must always follow the order of operations:

Brackets → Indices → Division/Multiplication → Addition/Subtraction

Volume of a cylinder

The formula for calculating the volume of a cylinder is:

$V = \pi h r^2$ Where V = volume
r = radius h = height

A cylinder has a height of 10 cm and a radius of 2 cm.
Calculate the volume of the cylinder.
$V = \pi \times 10 \times 2^2$
$V = \pi \times 10 \times 4$
$V = 125.6637061 \text{ cm}^3$

Perimeter of a rectangle

The formula for calculating the perimeter of a rectangle is:

$P = 2 \times (l + w)$ Where P = perimeter
l = length w = width

A rectangle has a perimeter of 20 cm and a width of 4 cm.
Calculate the length of the rectangle.
$P = 2 \times (l + w)$
$20 = 2 \times (l + 4)$
$20 = 2 \times l + 8$
$12 = 2 \times l$
$l = 6 \text{ cm}$

Formulae and equations

Equations have an equal sign: what is on the left equals what is on the right. A formula is a type of **equation** that has more than one variable.
Formulae describe the relationship between different variables.

For example:
$x - 7 = 12$ is not a formula because it only has one feature.
$A = \pi r^2$ is a formula, it has two features A and r. Remember that π is a number.

In this section you will:
- ◉ choose appropriate values and **variables**
- ◉ substitute into formulae
- ◉ use the correct order of operations
- ◉ show your method in a clear and concise way

 Take a look: Taxi ride

A taxi company uses this formula when charging clients.

$C = 1.6 + 1.2\,m + 0.13\,w$ C = cost of taxi ride (£)
                                     m = miles travelled
                                     w = minutes spent waiting

Josh has to wait ten minutes for a client.
He then drives the client four miles.

Q How much should he charge the client?

💡 Here's a possible solution:

$m = 4$ and $w = 10$

> **R** Select the information from the question and say what it is equal to

$C = 1.6 + 1.2\,m + 0.13\,w$
$C = 1.6 + 1.2(4) + 0.13(10)$
$C = 1.6 + 4.8 + 1.3$
$C = £7.70$

> **I** Use a set of brackets to show that you have substituted in

 **Have a go**

1 The formula for converting Fahrenheit to Celsius is:
   $C = \frac{5}{9}(F - 32)$ C = Celsius
                                 F = Fahrenheit

Q How many degrees Celsius is 90°F?

2 When you put money into a bank account the bank will usually pay you
   interest. You can calculate simple interest using the formula below.

   simple interest = amount of money × time in years × interest rate ÷ 100

Q Paul puts £435 in the bank on 1 March. The interest rate is 2.4%
   How much interest can Paul expect by 31 August?

3 Cara works in the sales team of a stationery company.
Her wages are calculated using the formula below.

$$P = 8(b + 2h) + \frac{s}{20}$$

P = total pay (£)
b = basic number of hours worked
h = hours of overtime worked
s = amount of stationery sold (£)

Q In January, Cara worked a total of 145 hours. 25 hours of the 145 hours
were overtime. She sold a total of £4200 of stationery.
How much should Cara be paid?

We're on the way

In this section you will:
- choose appropriate values from tables
- recognise when values need to be changed
- use mathematical techniques to change values
- explain the differences between two sets of data
- decide if the results are appropriate

Take a look: Calculating BMI

Doctors use Body Mass Index (BMI) as a measure of obesity.
The formula for calculating BMI is:

$$B = \frac{w}{h^2}$$

B = BMI (kg/m^2)
w = weight (kg)
h = height (m)

Category	BMI range (kg/m^2)
underweight	from 15 to 18.4
normal	from 18.5 to 22.9
overweight	from 23 to 27.5
obese	from 27.6 to 40
morbidly obese	greater than 40

Nathan and Karim want to know whose weight is most appropriate for
their height.
Nathan has a height of 160 cm. He weighs 45 kg.
Karim has a height of 170 cm. He weighs 55 kg.

Q Use the formula and table to compare the BMI of Nathan and the
BMI of Karim.

Here's a possible solution:

Nathan's height = 160 cm = 1.6 m
Karim's height = 170 cm = 1.7 m

> **A** The formula needs you to convert the heights into metres

Nathan's BMI = $45 \div 1.6^2$ = 17.578125
Karim's BMI = $55 \div 1.7^2$ = 19.031141

> **!** Use these results to choose the appropriate category from the table

Nathan is classified as underweight and Karim is classified as normal. So Karim has a weight most appropriate for his height.

> **!** Consider if the BMIs are appropriate

 ## Take a look: Slimming club

A slimming club uses a points system to help people lose weight. Each serving of food is scored using the formula below.

$$P = \frac{e}{50} + \frac{f}{12} - \frac{d}{5}$$

P = number of points
e = energy (kcal)
f = fat (g)
d = fibre (g)

Sugar-free orange squash	
Typical values per 100 ml	
Energy	125 kJ
	30 kcal
Protein	0 g
Carbohydrates	5.5 g
of which **sugars**	5.5 g
Fat	0 g
Fibre	0 g
Salt	0 g

Fresh orange juice	
Typical values per 100 ml	
Energy	184 kJ
	43 kcal
Protein	0.6 g
Carbohydrates	10.0 g
of which **sugars**	10.0 g
Fat	0.1 g
Fibre	0 g
Salt	0 g

Oliver uses too many points each day so he decides to drink sugar-free orange squash rather than fresh orange juice.

To make the squash, Oliver mixes water and orange cordial in a ratio of 4:1

 Oliver was drinking an average of 750 ml fresh orange juice each day.
Now he drinks an average of 750 ml orange squash.
How many points will he save?

Here's a possible solution:

First calculate the points in 750 ml of fresh orange juice:

$$e = 43 \times 7.5 = 322.5 \text{ kcal}$$
$$f = 0.1 \times 7.5 = 0.75 \text{ g}$$
$$d = 0$$

> **R** Use the nutritional information tables to find the values for substituting into the formula

$$P = \frac{322.5}{50} + \frac{0.75}{12} - \frac{0}{5}$$
$$P = 6.45 + 0.0625 - 0$$

> **A** Convert 100 ml into 750 ml

$$P = 6.5125$$

The fresh juice is worth 6.5125 points.

Now calculate the points in 750 ml of sugar-free orange squash:

$$750 \div 5 = 150 \text{ ml of cordial}$$
$$e = 30 \times 1.5 = 45 \text{ kcal}$$
$$f = 0 \text{ g}$$
$$d = 0 \text{ g}$$

> **A** Use the ratio to find the amount of cordial needed to make 750 ml of squash

$$P = \frac{45}{50} + \frac{0}{12} - \frac{0}{5}$$
$$P = 0.9 + 0 - 0$$
$$P = 0.9$$

> **Fact**
> Water is not worth any points.

The sugar-free orange squash is worth 0.9 points. The difference in points is:

$$6.5 - 0.9 = 5.6 \text{ points}$$

Oliver will save 5.6 points a day if he stops drinking fresh orange juice and drinks sugar-free orange squash.

 Have a go

Milk 1 cup (244 ml)	
Typical values	
Energy	511 kJ
	122 kcal
Protein	8.1 g
Carbohydrates	12.5 g
Fat	4.9 g
Fibre	0.0 g

Cocoa powder 3 teaspoons	
Typical values	
Energy	466 kJ
	111 kcal
Protein	1.9 g
Carbohydrates	23.4 g
Fat	1.2 g
Fibre	1.0 g

4 The formula for calculating weight loss points is given on page 69.

(Q) How many points are in 150 ml of milk?

5 Kanicka mixes cocoa powder and hot milk in the ratio 1:12 to make 260 ml of hot chocolate. One teaspoon of cocoa powder has a capacity of 10 ml.

(Q) Use the formula on page 69 to calculate the points value of Kanicka's hot chocolate.

> **Fact**
> 1 inch = 2.54 cm

6 Rochelle weighs 90 kg. She is 5 feet 9 inches tall.
She wants to lose weight, so she starts a fitness programme.

Rochelle would like to have a BMI of 20 in four months' time.
The formula for calculating BMI is on page 68.
Rochelle wants to lose the weight at a steady rate.

> **Think First!**
> Remember that Rochelle's height will remain the same.

(Q) Design a calendar that gives Rochelle a weight to aim for each month.

Exam ready!

In this section you will:
- decide on a logical mathematical process
- use a variety of inputs to analyse the effect on the final solution
- write results to an appropriate degree of accuracy
- advise on a number of different outcomes
- draw conclusions and justify your solutions

Take a look: Insulation

The construction industry uses R-values to measure the insulating properties of materials.
When materials are mixed in layers, their R-values can be added to calculate a total R-value.
The formula for calculating R-values is:

$$R = \frac{dat}{h}$$

R = R-value
d = temperature difference (Fahrenheit)
a = area (square feet)
t = time (hours)
h = heat loss (Btu)

> **Fact**
> Btu stands for British thermal unit.

Insulation materials	R-value for 1 inch
Loose fill	2.08
Foam	5.3
Blown loose fill	3.6

Damien is designing the walls for a building.

The insulation for the walls needs to be less than 4.5 inches thick.
It must have a heat loss of less than 50 000 Btu a day.
The surface area of the outside wall is 1300 feet².
The average temperature difference is 20°F.

(Q) **Advise Damien on which of the three types of insulation materials he can use.**

(💡) Here's a possible solution:

$d = 20°F$
$a = 1300\ ft^2$
$t = 24$ hours
$h = 50\ 000$ Btu

> (R) Start by finding the minimum required R-value of the wall

Required R-value = 20 × 1300 × 24 ÷ 50 000 = 12.48

Loose fill = 12.48 ÷ 2.08 = 6 inches
Foam = 12.48 ÷ 5.3 = 2.35471698 inches
Blown loose fill = 12.48 ÷ 3.6 = 3.466666 inches

> (R) Calculate the depth needed for the different types of insulation

Damien can use blown loose fill or foam.

> (I) The loose fill would be thicker than 4.5 inches

 **Have a go**

7   A service engineer uses the formula below to calculate how much to charge each client.

   Total charge (£) = number of hours × 30 + cost of parts including VAT (£) + £30

He arrives at a client's house to fix their washing machine at 10:32 am. He leaves at 1:47 pm.
He uses a new drum costing £87, excluding VAT. He also uses three
sets of bolts costing £1.45 per pair, excluding VAT.
VAT is charged at a rate of 17.5%.

(Q) **How much does the service engineer charge the client?**

8   The table below gives the R-values for different types of windows.

Window	R-value
single glass	0.8
double glass	1.69
triple glass	2.56

All of the windows in Sarah's house are made out of single glass.

Sarah's house has two sizes of window. The small windows are 2 feet wide and 3 feet 3 inches high. The large windows are 2 feet 6 inches high and 4 feet wide.

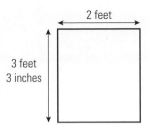

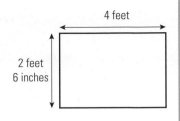

Q **a)** Calculate the area of each window in square feet.

Sarah wants to work out the reduction in heat loss for an average temperature difference of 10 °F and a time of 1 hour.

b) Use the R-value formula on page 71 to calculate the reduction in heat loss when:
   i) a small window is replaced with double glass
   ii) a large window is replaced with triple glass

9 John is building the foundations for a shed out of concrete. The foundations form a cuboid. The cuboid is 1 foot deep, 5 feet wide and 6 feet long.

Q **a)** Calculate the volume of the foundations in cubic feet.

To make the concrete he needs to mix aggregate, cement and water in a ratio of 7:2:1
The ratio refers to the volume of each material.

Q **b)** Calculate the volumes of cement, aggregate and water needed.

Concrete is made out of cement, aggregate and water.
Builders use the formula below to calculate the weight needed for each material.

$$V = \frac{w}{28.4G_s}$$

V = volume (cubic feet)
w = weight of loose material (kg)
G_s = Specific gravity

Material	Specific gravity
cement	3.2
aggregate	2.6
water	1

Q **c)** Use the formula above to calculate the weight of cement and the weight of aggregate needed.

Now you can:

- Use unfamiliar formulae to solve practical problems
- Choose appropriate information from tables and lists
- Substitute into **algebraic formulae**
- Make conclusions and compare results

8 Area, perimeter and volume
Know Zone

Area

The **area** of a shape is a measure of the space it takes up.

Rectangle	Triangle	Parallelogram	Trapezium
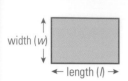	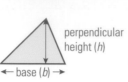	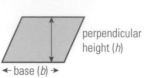	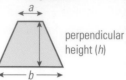
Area = length × width $A = lw$	Area = $\frac{1}{2}$ × base × height $A = \frac{1}{2}bh$	Area = base × height $A = bh$	Area = $\frac{1}{2}(a + b)h$

Perimeter

The distance around the outside of a shape is called the **perimeter**.

Perimeter
= 3 cm × 6 = 18 cm

Circles

The perimeter of a circle is called the **circumference**.

You need to know the formulae for the area and circumference of a circle.

Area = π × radius × radius
$$A = \pi r^2$$

Circumference = π × diameter
$$C = \pi d$$

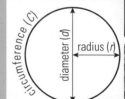

Volume

A **prism** is a 3-dimensional (3D) shape with a constant cross-section. The amount of space a 3D object uses is called the **volume**.

You can calculate the volume of a prism by multiplying the area of the cross-section by the length of the shape.

Area of cross-section
$A = \frac{1}{2}bh$
$A = \frac{1}{2} \times 8 \times 6$
$A = 24\ cm^2$

Volume of prism
$V =$ Area of cross-section × length
$V = 24 \times 12$
$V = 288\ cm^3$

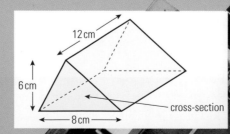

In this section you will:
- ◉ choose appropriate lengths
- ◉ make sure all measurements are in the same units
- ◉ use the formulae for the area of a rectangle and the volume of a cuboid
- ◉ decide if you should round solutions up or down

🔍 Take a look: Post Office charges

The Post Office classifies packages according to their size.

Letters	Large letters	Packets
Length: 24 cm max	Length: 35.3 cm max	Length: 61 cm max
Width: 16.5 cm max	Width: 25 cm max	Width: 46 cm max
Thickness: 5 mm max	Thickness: 25 mm max	Depth: 46 cm max

Q Calculate the maximum volume of each of the three types of package.

💡 Here's a possible solution:

Letters = $24 \times 16.5 \times 0.5 = 198 \, cm^3$ ————————— **R** Convert mm into cm

Large letters = $35.3 \times 25 \times 2.5 = 2206.25 \, cm^3$

Packets = $61 \times 46 \times 46 = 129\,076 \, cm^3$

⊕ Have a go

1 Nikhil has invited 30 guests to his birthday party. He makes fruit cocktail in a container shaped like a cuboid. The container has a width of 25 cm, a height of 36 cm and a length of 42 cm.

Nikhil will use equal amounts of orange, pineapple and cranberry juice.
Each of the three juices is sold in 1-litre carton. Nikhil will use all of the juice in each carton he buys.

Fact

1 litre = 1000 cm³

36 cm
42 cm
25 cm

Q What is the maximum number of orange juice cartons Nikhil needs to make enough cocktail to fill his container?

Think First!

Should the number of juice cartons be rounded up or down?

2 Ben is designing his garden. He wants a rectangular patch of grass with a width of 6 m and a length of 8 m.

Ben goes to the local garden centre. He finds boxes of grass seed on sale for £4.99 each.

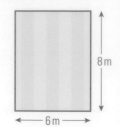

(Q) Each box of seed will cover an area of 10 m². How many boxes of seed will Ben need?

3 Chris and Becky are holding a wedding reception.
There will be 12 circular tables.
Each table has a diameter of 2 m.

Chris and Becky decide to put ribbon round the edge of each table.
The ribbon costs £1.20 per metre.
Ribbon is sold in whole-metre lengths (for example, Chris and Becky could buy 5 m or 6 m of ribbon, but not 5.5 m of ribbon).
You can calculate the circumference of a circle using the formula: $C = \pi d$

(Q) Calculate the total cost of the ribbon.

● We're on the way

In this section you will:
- calculate the area of compound shapes
- communicate facts and figures when you give outcomes

Take a look: Office space

The table shows the standards used to estimate the amount of floor space needed for a business.

Type of space	Area needed
Chairman's office	30 m²
Executive's office	12 m²
Supervisor (open space)	9 m²
Secretary (open space)	7 m²
Reception area	19 m²
Lunch room	1.2 m² (per person, not including kitchen) kitchen should be $\frac{1}{3}$ of seating area
Corridors	20% of total space

A company called Prime Solutions employs one chairman, three executives, ten supervisors and four secretaries. The company needs an office space with a reception area and a lunch room with a kitchen.

The floor plan for an office is shown below.

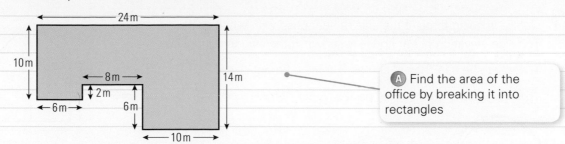

(A) Find the area of the office by breaking it into rectangles

(Q) Does the building have the total floor space that Prime Solutions needs?

(💡) Here's a possible solution:

Chairman's office = 30 m²
Executive offices = 3 × 12 = 36 m²
Supervisor's space = 10 × 9 = 90 m²
Secretary's space = 4 × 7 = 28 m²
Reception area = 19 m²
Lunch room (without kitchen) = 1.2 × (1 + 3 + 10 + 4) = 21.6 m²
Kitchen = 21.6 ÷ 3 = 7.2 m²

Total space needed (without corridors) = 30 + 36 + 90 + 28 + 19 +
                                          21.6 + 7.2 = 231.8 m²

Total space needed (including corridors) = 231.8 × 1.2 = 278.16 m²

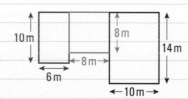

(❗) When you give advice, come to a conclusion and state the relevant measurements

Total available space = red area + blue area + green area
                     = (10 × 14) + (6 × 10) + (8 × 8)
                     = 140 + 60 + 64
                     = 264 m²

The office doesn't have the space needed. Prime solutions needs a bit more than 278 m², but the building is only 264 m² in size.

 **Have a go**

4 Natalie is going to paint the side of her house with white paint.
The side elevation of her house is shown in the diagram opposite.

The paint she has chosen comes in three different sizes.

The instructions on the tin of paint say that 1 litre will cover an area of 5 m².

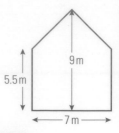

Think First!

Don't forget to divide by two to find the area of a triangle.

Q What is the minimum cost of the paint Natalie needs?

5 A council has paid £80 000 for a plot of land to build a car park.

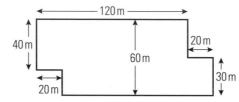

Think First!

Think about different ways of finding the area.

The area needed for each parking space is 28 m². This area includes the circulation areas needed for people to drive their cars in and out.

The council estimates that they will get £8.50 a day for each space.
It will cost £150 000 to build the car park.

Q Advise the council on how long it will take to get back the amount of money they spent on buying the land and building the car park.

6 When natural gas is transported it is converted into liquefied natural gas (LNG).
LNG takes up $\frac{1}{600}$ the volume of natural gas.
LNG is stored at low temperatures.

When LNG is transported by sea, an estimated 5.2% of the LNG is lost due to evaporation.

LNG fills the container shown on the right.

The price of natural gas is £2.50 per cubic metre.

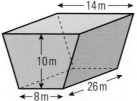

Q How much money will be lost due to LNG evaporating from this container?

Exam ready!

In this section you will:

- ◉ read and follow sets of instructions
- ◉ use diagrams to model a situation
- ◉ think about different solutions to a problem
- ◉ advise on the advantages and disadvantages of different options

Take a look: Buying carpet

Mahmuda and Farhad want to put a carpet on their living room floor.
The carpet they have chosen costs £20 per square metre.

The carpet is sold in 5 m widths.
Mahmuda and Farhad can buy any length of the carpet.

They want the minimum length of joins in the carpet.

Q Advise Mahmuda and Farhad on the best way to put the carpet on the floor.

💡 Here's a possible solution:

| **A** Think about different ways of putting the carpet on the floor | **R** Use diagrams to show how the carpet will be put on the floor | **R** Use diagrams to show the carpet before it is cut |

Option A

Total area = 5 × 9 = 45 m²

Total cost = 45 × 20 = £900

Length of join = 8 − 4.5 = 3.5 m

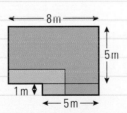

Option B

Total area = 5 × 9 = 45 m²

Total cost = 45 × 20 = £900

Length of join = 3 m

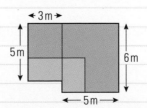
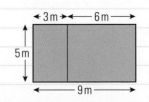

Option A and Option B both cost £900.
The join in Option B is 0.5 m shorter than the join in Option A.

Option B is best for Mahmuda and Farhad because they want the minimum length of joins.

7 A Sea Life centre has a new aquarium for keeping ocean fish.
The aquarium is a cylinder with a height of 3 m and a radius of 4 m.

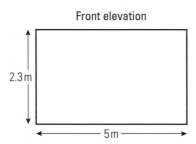

You can calculate the area of a circle using the formula:

$A = \pi r^2$

The owners of the aquarium need to add salt to the water before they can put fish in the aquarium.

They need to add 35 g of salt for every litre of sea water.

Q **a)** Work out the volume of the aquarium.

b) How much salt is needed?

> **Fact**
> $1 \, m^3 = 1000$ litres

8 Carey is going to wallpaper one wall in her bedroom.

The wall is 5 m wide and 2.3 m high.

Front elevation

```
2.3 m |                    |
       |                    |
       |_____|
            ← 5 m →
```

Carey finds this advice about putting up wallpaper.

Carey has chosen a wallpaper in 45 cm × 5 m rolls.

> **Five steps to hanging wallpaper**
> 1 Strip walls
> 2 Measure the length of paper needed. This is usually the height of the room with an additional 50 mm at each end
> 3 Put wallpaper paste on the back of the paper
> 4 Hang the paper vertically and cut off excess wallpaper at the top and bottom
> 5 Push out any air bubbles
> 6 Repeat until the wall is covered

Q **a)** How many rolls of wallpaper does Carey need to buy?

Each roll costs £14.99.
A pot of wallpaper paste covers approximately 5 m²
of wallpaper. It costs £6.99

Q **b)** How much should Carey budget for spending on wallpaper and paste in total?

9 Javed wants to build a house.

The floor plans of the house are shown below.

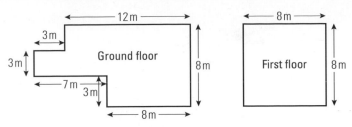

Q **a)** Work out the area of the ground floor and the area of the first floor.

The table gives the cost in £ per square metre of total internal floor space for building a new house for three types of builder.

Total internal floor space = area of ground floor + area of first floor

		Type of builder		
		DIY and subcontractors cost in £	Subcontractors cost in £	Main contractor cost in £
Double storey	Small (90–130 m²)	865	916	1017
	Medium (131–220 m²)	728	771	857
	Large (221 m²+)	672	712	791

Key. 898 = £898/m²

Q **b)** Javed wants to use subcontractors to do all the work.

How much will building the house cost Javed?

Now you can:

- Calculate areas and volumes of real life objects
- Use appropriate units of measurement
- Choose relevant information from tables and written instructions
- Give solutions to an appropriate degree of accuracy
- Communicate different solutions to problems
- Think about the advantages and disadvantages of different solutions

In the Functional Skills Mathematics exam there will be questions where you need to communicate decisions and arrive at solutions to practical problems. In many of these questions it is a good idea to think about how you will work out your answer before you start.

ResultsPlus
Maximise your marks

A charity wants to send tins of food to a foreign country needing food aid. The charity will pack the tins into boxes. There will be 60 tins in each box.

The height of each tin is 10.5 cm.
The diameter of each tin is 7 cm.
Each box is a rectangular cuboid.

There must be fewer than eight tins along any edge of the box.

Work out the dimensions of a suitable box for 60 tins of food.

FOOD AID   10.5 cm

7 cm

Exam tip

To find the dimensions of the box think about three numbers that multiply together to equal 60. Each number must be less than 8.

Student response	Exam comments
Let's look at a poor answer: $7 \times 10.5 = 73.5 \text{ cm}^3$ $6 \times 10 = 60$ $6 \times 10.5 = 63 \text{ cm}$ $6 \times 7 = 42 \text{ cm}$ $10 \times 10.5 = 105 \text{ cm}$ The box would need to be 105 cm tall so that 60 tins can be stacked upright.	This student has begun the process of finding three numbers that multiply together to equal 60. The student has chosen 6, 6 and 10 as the three numbers but there can be no more than 8 tins along any edge of the box. The student gives incorrect box dimensions.
Let's look at an answer that will gain some marks: The bottom of the box is 6 × 5. Tin: $h = 10.5, d = 7$ Volume = $3.14 \times 10.5 \times 7 \times 7$	This student has worked out the dimensions of the base of a box that would lead to a correct answer. They then try to find the volume of a tin, but this is **not** what the question asks for.
Let's look at a complete answer: 30 tins × 2 = 60 15 × 2 × 2 6 × 5 × 2 6 by 5 by 2 (6 along 5 along 2 up) 42 cm along by 35 cm along by 21 cm up	This student has found the correct number of tins along each side of the box. The dimensions of the box are written in a clear way.

Exam question

Alisha is going to put tiles on part of a bathroom wall.
The tiles are **5 cm square.**
The tiles will cover a rectangular space with dimensions 1.25 m by 0.75 m.

Plan of the rectangular space

The scale of the plan is 1:10
One colour of tile will cover the shaded outer part.
Two different colours of tile will cover the inner
part in the ratio 1:2

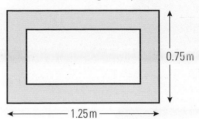

Colour of tile
Terracotta
Buttermilk
Blue

**Decide the number of tiles of each colour Alisha would need for the design.
Show clearly how you get your answer.**

March 2010 (8 marks)

How students answered

Students chose two main approaches:
- Finding the area of wall to be covered and the area of the tiles
- Filling the different parts of the wall space with tiles

Marks were awarded for progress in these three areas:
- Using lengths and working with areas
- Thinking about how many tiles will fill the space in the diagram
- Understanding how to use ratio to produce an answer to the problem

■ 60% of students (0–2 marks)	● 22% of students (3–5 marks)	▲ 18% of students (6–8 marks)
Students who worked out or measured a length from the diagram gained marks.	Students who measured a length, then found an area or the number of tiles needed to fill the wall space were awarded more marks	Students who correctly found the number of tiles for the shaded part of the wall and began to use ratio to find the number of tiles needed for the inner part gained more marks. Students who successfully used ratio to find the number of tiles needed for all the wall and stated the colours of the tiles gained full marks.

Put it into practice

1 Find some cardboard boxes and measure each side. Think about how many tins of different sizes can be placed in the boxes.

2 Find different sized tiles and work out how many tiles you would need to cover a particular wall.

ASSESSMENT PRACTICE 2

1 A man has complained to his local council. He says that a neighbour's hedge is causing loss of light to his garden.
If the hedge is too high, the council can make the neighbour cut it down.

A plan view of the man's house and garden is drawn below.

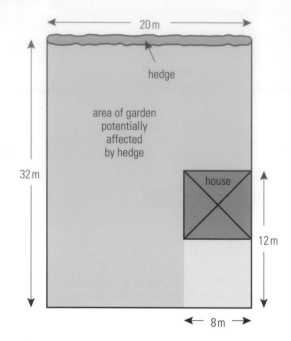

The council uses the following formula to work out the maximum height allowed for a hedge:

$$H = \frac{0.35A}{L}$$

H = the maximum height of hedge allowed
A = the area of the garden that might be affected by the hedge
L = the length of the hedge

The height of the hedge is currently 8 m.

(Q) Should the council order the neighbour to cut the hedge?

2 A teacher is going to order some exercise books ready for the new school year.
She will store the exercise books in a shelving space of 1.5 m wide by 1 m deep by 0.8 m high.

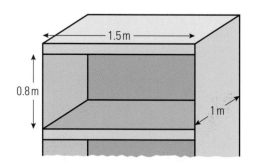

The exercise books are sold in packs of 25.
For safety reasons, the books must be stored flat.
They cannot be stored in piles of more than five packs.

(Q) a) Work out the height, in cm, of five packs of exercise books.

The teacher must store all the exercise books in the shelving space.
It is cheaper to buy the exercise books in large quantities.
She wants to buy as many exercise books as possible.

(Q) b) How many packs of exercise books should the teacher buy?

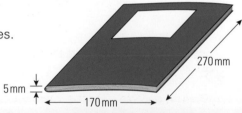

3 Anoop wants to buy some chickens.
He has seen three different chicken coops advertised on the Internet.

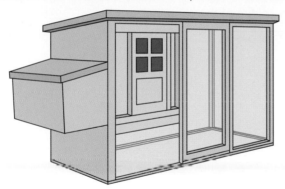

Happy Hutch: Dimensions of 90 cm long by
                        88 cm wide by 79 cm high

Poultry Ark: Dimensions of 895 mm long by
                        850 mm deep by 910 mm high

Chicken Cottage Coop: Dimensions of 71 inches long
                        by 30 inches wide by 39.5 inches high

1 inch = 2.54 cm   12 inches = 1 foot

1 foot = 30.48 cm 144 inches^2 = 1 square foot

The table shows the area of ground needed for one chicken of three
different breeds.

Breed	Area in square feet
very small	1
medium	1.5
large	2

Anoop wants to buy the chicken coop with the largest floor area.

(Q) a) Which chicken coop should Anoop buy?

Anoop decides to buy 3 large chickens and then as many medium
chickens as possible.

(Q) b) How many medium chickens should Anoop buy?

4 A manufacturer makes boxes.
Each box is a rectangular cuboid with
dimensions of 9 cm by 21 cm by 30 cm.

The manufacturer wants to pack the boxes
into cartons to transport them.
Each carton must be large enough to contain 24 boxes.

9 cm

21 cm   *(not drawn to scale)*

30 cm

(Q) Design a carton for the manufactureer to use.

5 The diagram shows the net for an open container.
The container will be used to hold noodles.
Fastening clamps will be fixed to the sides of the container.

(Diagram is drawn to scale)

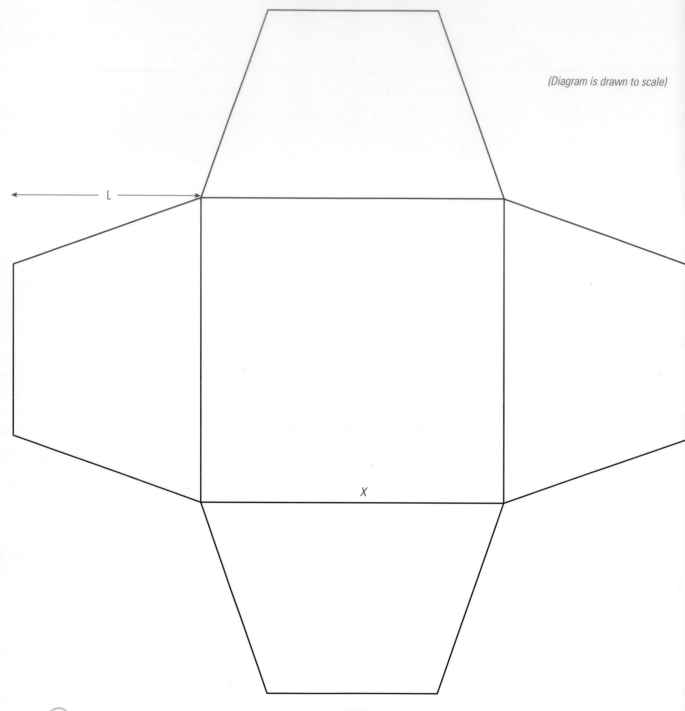

Q **a)** Measure the lengths X and L on the diagram.

The scale of the net to the container is 1:2

Q **b)** Work out the lengths X and L on the actual container.

The formula used to find the total surface area of the container is:

$S = X^2 + 2L(X + 9)$

S = total surface area of the container (cm²)

L = height of a trapezium (cm)

X = length of the square base of the container (cm)

The amount charged by the manufacturer to make the containers depends on the surface area of the net

(Q) c) Work out the total surface area of the container.

---

6 Nesta has taken the carpet out of her living room. She wants to cover the floorboards with varnish or paint.

Nesta will need to put 2–4 coats of varnish or 2–4 coats of paint on the floorboards.

A plan view of Nesta's rectangular living room is drawn below.

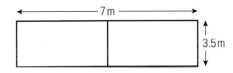

Nesta needs to choose between floor varnish and floor paint.

Floor varnish: 2.5 litres covers 40–50 square metres
A 2.5-litre tin costs £33.99

Floor paint: A 1-litre tin covers 10 square metres. It costs £7.90
A 5-litre tin covers 50 square metres. It costs £29.95

(Q) Compare the cost of using floor varnish and floor paint.

---

7 A jewellery manufacturer is going to make a new design of bracelet. The manufacturer will make the bracelet from soft metal pins.

The manufacturer will use two different sized pins.
Pin 1 is a square with side length 5 mm.
Pin 2 is a square with side length 8 mm.

The length of the bracelet will be between 230 mm and 240 mm.
There will be a clasp at the end of the bracelet.
Each clasp is a square with side length 8 mm.

The design of the bracelet is shown in the diagram below.

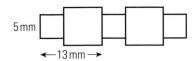

(Q) How many pins are needed to make the bracelet?
Explain your answer.

Tables, **graphs** and charts are all around us. We see them in advertisements, holiday brochures, magazines and newspapers.

Tables

To extract information from a **table** or spreadsheet, find the correct row and column and then move along and down to find the figure you want.

In this holiday brochure the cost of a 14-night stay for one adult leaving on 1 June is £739.

Accomodation	Sunset Holiday Village			Child Prices for Sunset	
Board basis	All Inclusive				
Price based on	Twin Adult			1st Child	2nd Child
No. of nights	7	14	Ex Day	Any duration	
01 May–14 May	459	629	28	364	276
15 May–21 May	474	689	28	379	284
22 May–28 May	544	729	28	449	359
29 May–09 Jul	539	739	33	424	314
10 Jul–23 Jul	639	964	46	489	334

Bar charts, pie charts, line graphs

You will also need to extract information from **bar charts**, **pie charts** and **line graphs**.

A dual bar chart can be used to compare two sets of **data**. The **dual bar chart** below shows the average midday temperatures in Milan and Naples during the summer months.

When you read a bar chart, make sure that you look carefully at the scale and work out what any small subdivision is worth.

The difference in temperature between Milan and Naples in May is estimated to be 4°C.

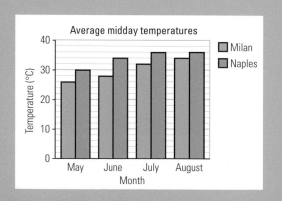

Scatter graphs

You will need to know how to draw conclusions from **scatter graphs**.

Scatter graphs can be used to show relationships between two sets of data. As one value increases the other value may increase or decrease.

The data in this scatter graph suggests that students who watch more TV generally achieve lower GCSE results.

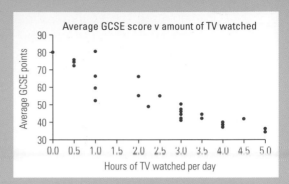

You need to be able to draw tables, spreadsheets, bar charts, pie charts and line graphs and label them correctly.

When you draw a graph choose sensible scales and give the graph a title.

Line graphs

In an experiment boiling water is poured into a cup. The temperature of the water is measured in degrees Celsius every 2 minutes.

The line graph shows the results of the experiment.

The axes are labelled with sensible scales and the graph has been given a title.

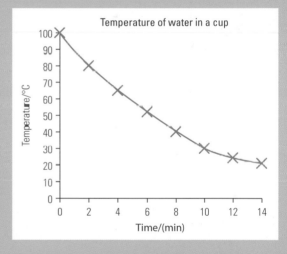

Let's get started

In this section you will:
- learn to recognise redundant information
- understand which is the most appropriate graph to use for particular kinds of data
- read axes and scales accurately
- communicate results clearly giving reasons for your conclusions

Take a look: Stopping distances

The graph shows the stopping distances (in metres) of a car when it is travelling at different speeds (in miles per hour).

It gives the stopping distances for both poor and good conditions.

Q A car is travelling 100 m behind the car in front.
What would be a safe driving speed?

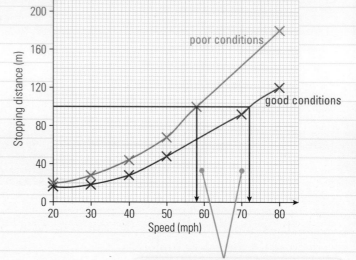

Here's a possible solution:

In poor conditions a safe driving speed would be up to 58 mph. In good conditions the car could safely be travelling at up to 70 mph (the maximum speed limit).

A Draw a line on the graph to show a stopping distance of 100 m. Use the line to find the maximum speed that a car should be travelling to be able to stop safely

Have a go

1 The teacher of an exercise class measured the weight and the pulse rate of 12 people in the class.
She put the data on a scatter graph.

Q Comment on the results.

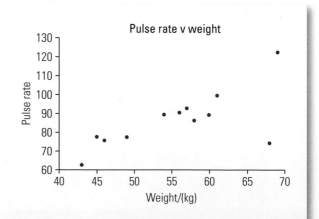

2 John says that all we need is food to keep us alive, clothes to keep us warm and housing for shelter. All other things are luxuries.

Q Compare the spending habits of a family in 1957 with a family in 2006. From the data, in which year would John suggest the family has a more luxurious lifestyle?

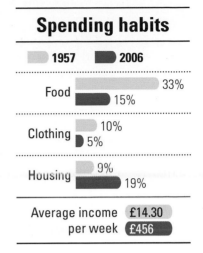

Spending habits

1957   2006

Food   33%   15%

Clothing   10%   5%

Housing   9%   19%

Average income £14.30 per week £456

3 Michael and Sarah are going on their honeymoon for 14 days in the summer. They are going to choose between the resorts of Blaine and Carnet.

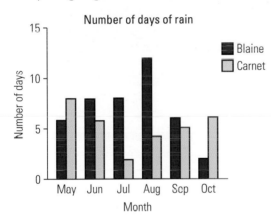

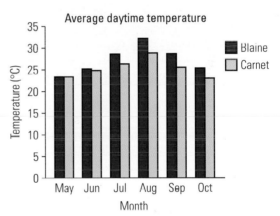

Q Use the graphs to advise Michael and Sarah of the resort to choose. Give full reasons for your choice.

4 Sharini manages a company called Thermoplastics. The company has been trading since 2006. The company's profits for the past 4 years are shown in the table below.

	2006	2007	2008	2009
Profit	£67 000	£130 000	£140 000	£48 000

Sharini writes a report to the people who have shares in the company every year.

Q Draw an appropriate graph that Sharini could use to describe the trend in profits over the last four years.
Write a short paragraph for Sharni to use with the graph.

We're on the way

In this section you will:
- understand how to find information from pie charts
- analyse and understand the shape of a line graph
- explain relationships and make conclusions

Take a look: School council election

There is an election for a student representative on the school council.
There are four candidates, James, Kayleigh, Emma and Habib.

The two pie charts show the votes cast for each candidate.

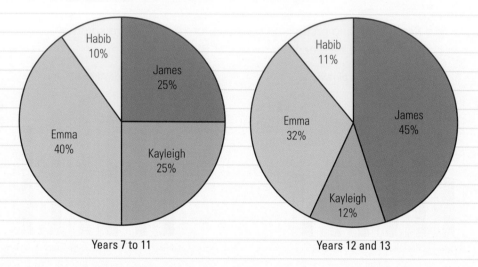

| Years 7 to 11 | Years 12 and 13 |

1000 students from Years 7 to 11 voted.
300 students from Years 12 and 13 voted.

Q Which candidate won the election?
Explain why.
Give full reasons for your answer.

Here's a possible solution:

James gets 45% of 300 + 25% of 1000
= 135 + 250 = 385 votes
Emma gets 32% of 300 + 40% of 1000
= 96 + 400 = 496 votes

R You cannot use the percentages on the pie chart because a different number of students in each age group voted

Therefore, Emma won the election.

Have a go

5 The sales figures for a car and commercial vehicle company are shown below.

Year	2005	2006	2007	2008	2009	2010
Car sales	53	45	76	85	105	115
Commercial sales	26	22	35	43	12	14

(Q) Draw a suitable chart or graph to display the data.
Identify any trends.

6 Karen wants to join a gym. She can choose Alpha or Bodies or Curves.
The graph shows the charges of each gym against the number of sessions people attend per year.

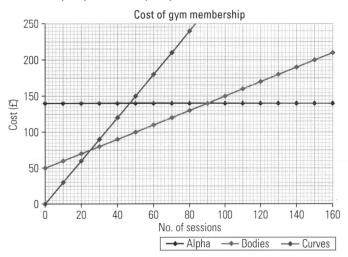

(A) The line representing the Alpha gym is horizontal because you pay a fixed fee of £140

(Q) Advise Karen on the gym to join.

7 Gordon works for an event management company.
The company has a wedding reception to organise for 300 guests.
The guests will be seated at 50 tables.
Gordon has to design a recording sheet so that each of the 10 waitresses can take meal orders from guests at the tables they serve.

(Q) Design your own recording sheet to collect the information as efficiently as possible.
The menu for the event is shown.

MENU

Starters
Choice of:
• Cream of mushroom soup
• Pate with toast

Main course
Choice of:
• Fillet of beef
• Roast chicken
• Vegetarian Option

Sweets
Choice of:
• Death by Chocolate
• Ice Cream and Strawberries

Think First!

Think about what is important here. The waitresses need to know the meals each table of guests ordered. They do **not** need to know the meal each **person** ordered.

8 An article in a newspaper stated, 'Female sprinters will be able to compete against male sprinters and beat male sprinters by the middle of the 21st century!'

(Q) The data in the table shows the times of male and female Olympic champion sprinters over the past 60 years. Use the data to investigate if the statement is reliable or not.

Year	Gold Men	Country	Time (sec)	Gold Women	Country	Time (sec)
1948	Harrison Dillard	USA	10.3	Fanny Blankers-Koen	NED	11.9
1952	Lindy Remigino	USA	10.4	Marjorie Jackson	AUS	11.5
1956	Bobby Morrow	USA	10.5	Betty Cuthbert	AUS	11.5
1960	Armin Hary	GER	10.2	Wilma Rudolph	USA	11.0
1964	Robert Hayes	USA	10.0	Wyomia Tyus	USA	11.4
1968	James Hines	USA	9.9	Wyomia Tyus	USA	11.0
1972	Valeri Borzov	URS	10.14	Renate Stecher	E Germany	11.07
1976	Hasely Crawford	TRI	10.06	Annegret Richter	W Germany	11.08
1980	Allan Wells	GBR	10.25	Ludmilla Kondratyeva	USSR	11.06
1984	Carl Lewis	USA	9.99	Evelyn Ashford	USA	10.97
1988	Carl Lewis	USA	9.92	Florence Griffith-Joyner	USA	10.54
1992	Linford Christie	GBR	9.96	Gail Devers	USA	10.82
1996	Donovan Bailey	CAN	9.84	Gail Devers	USA	10.94
2000	Maurice Greene	USA	9.87	Marion Jones	USA	10.75
2004	Justin Gatlin	USA	9.85	Yulia Nesterenko	BLR	10.93
2008	Usain Bolt	JAM	9.69	Shelley Ann Fraser	JAM	10.73

(Q) Exam ready!

In this section you will:
- ⦿ choose a suitable method to answer the question
- ⦿ examine relationships by drawing line graphs or scatter diagrams
- ⦿ make conclusions and communicate results

Take a look: The Premiership

Paul states that teams with greater goal differences are placed higher in the league.

Goal difference = Goals for (GF) – Goals against (GA)

(Q) Use the Premiership league table opposite to test Paul's statement.

Position	Team	Pl	W	D	L	GF	GA	Pts
1	Manchester United	38	28	6	4	68	24	90
2	Liverpool	38	25	11	2	77	27	86
3	Chelsea	38	25	8	5	68	24	83
4	Arsenal	38	20	12	6	68	37	72
5	Everton	38	17	12	9	55	37	63
6	Aston Villa	38	17	11	10	54	48	62
7	Fulham	38	14	11	13	39	34	53
8	Tottenham Hotspur	38	14	9	15	44	45	51
9	West Ham United	38	14	9	15	42	45	51
10	Manchester City	38	15	5	18	58	50	50
11	Wigan Athletic	38	12	9	17	34	45	45
12	Stoke City	38	12	9	17	38	55	45
13	Bolton Wanderers	38	11	8	19	41	53	41
14	Portsmouth	38	10	11	17	38	57	41
15	Blackburn Rovers	38	10	11	17	40	60	41
16	Sunderland	38	9	9	20	34	54	36
17	Hull City	38	8	11	19	39	64	35
18	Newcastle United	38	7	13	18	40	59	34
19	Middlesbrough	38	7	11	20	28	56	32
20	West Bromwich Albion	38	8	8	22	36	67	32

 Here's a possible solution:

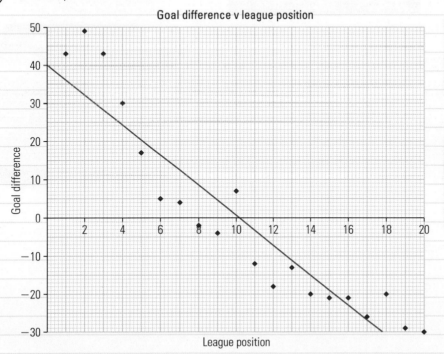

A Draw a scatter graph for goal difference against league position

R A scatter diagram is useful in making a link between league position and goal difference

The graph shows that the teams with the biggest goal difference are higher up the league. So Paul's statement is correct.

Have a go

9   Neal and Mandy have two children, David and Laura. The family are
    planning a holiday for three nights. David is 17 and Laura is 14.
    The family are going to Bramston's outdoor activity centre.

    They have £800 to spend.
    Each person wants to do at least two different activities.
    They can go in August or in October.

Accommodation (inc all meals)	High season (May to Sep)		Mid season (Apr and Oct)		Low season (Nov to Mar)	
Adult (17 and over) (per night)	£60		£50		£41	
Child (under 17) (per night)	£38		£30		£27	
Price per activity (10% reduction if pre-booked)	**Full day**	**Half day**	**Full day**	**Half day**	**Full day**	**Half day**
Adult (17 and over)	£30	£15	£25	£12.50	£20	£10
Child (under 17)	£18	£9	£18	£9	£15	£7.50

Q Plan the family holiday. Give full reasons for your decisions.

10 Leon wants to sell his car. He goes to the local garage to find cars of the
    same make and model.

Car	Age	Price (£)
A	6	4100
B	3	6500
C	8	2300
D	1	8530
E	2	6900
F	3	6700
G	5	6000
H	7	3050
I	2	7000
J	5	8250

ResultsPlus
Exam Tip

You need to draw a diagram
that will be appropriate.
In this case a scatter diagram
will be appropriate.

Leon's car is 4 years old.

Q Using the information given in the table, suggest a possible selling
price for his car.

11 Carol drives a car.
She wants to know the anual emissions CO_2 (kg) that her
car produces.

$$\text{Annual emissions } CO_2 \text{ (kg)} = \frac{\text{distance travelled (km)} \times \text{emissions rating (g/km)}}{1000}$$

Carol drives 20 000 km a year.
The emissions rating of her car is 205 g/km.

Q **a)** Use the formula to calculate the annual emissions CO_2 (kg)
    for Carol's car

Carol will replace the car with a new car.
She wants to reduce her annual emmisions by between 40 and 60%.
She wants to reduce her annual distance travelled by 5000 km.

The graph shows the emissions ratings for all the different types
of car Carol is thinking of buying.

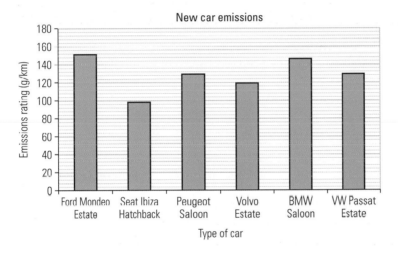

New car emissions

Q **b)** Use the information provided to make a new car choice
    for Carol.

Now you can:

- Understand how to use scales in charts and graphs
- Find relevant information from tables and graphs
- Use information from graphs and charts to solve problems
- Understand correlation and use it to predict outcomes

10 Use and interpret data
Know Zone

We hear the word 'average' used all of the time.

'She is about average height for her age.'

'The average house price went down by £10 000 last year.'

Mean, median, mode

There are three types of **average** that we can find for a set of data: the **mean**, the **median** and the **mode**.

In the first week of January last year, midday temperatures in London were:
0°C, 3°C, 1°C, 4°C, 2°C, 2°C and 2°C.

* To find the mean of the temperatures we add up all the temperatures.
  Then we divide the total by the number of temperatures: $14 \div 7 = 2°C$
* To find the median of the temperatures we arrange all the temperatures in order.
  Then we find the middle temperature: Median of 0, 1, 2, 2, 2, 3, 4 = 2°C
* The mode of the temperatures is the most common temperature.
  The mode is 2°C because there were three temperatures of 2°C

Each average can be useful in different situations.

	Most appropriate average	Least appropriate average
In 10 innings a cricketer scores: 12, 5, 6, 8, 10, 0, 11, 13, 100, 7	Median = 9 It is representative of most of the data	Mean = 172 ÷ 10 = 17.2 The score of 100 has a great effect on the mean. It is **not** representative because it is very different from the other scores.
The shoe sizes of seven friends are: 3, 4, 5, 5, 5, 5, 9	Mode = 5 Median = 5	Mean = 36 ÷ 7 = 5.2 No one has a shoe size of 5.2 because shoes are sold in whole or half sizes
In her maths tests this term, Jayne scores: 6, 6, 6, 6, 7, 7, 8, 8, 9, 10	Mean = 73 ÷ 10 = 7.3 Median = 7 (the middle value)	Mode = 6 6 is the most common score but it is also the lowest score

Range

We can also find the **range** of the temperatures above.

* To find the range we subtract the smallest value from the largest value: Range = 4 – 0 = 4°C

Comparing data

To compare two sets of data, work out the average and the range for both sets of data.

The table gives the mean height and range in heights of two Year 11 sports teams.
On average, the basketball team are taller.
The basketball team have a bigger spread of heights than the football team.

	Mean height (m)	Range in height (m)
Basketball team	1.78	0.21
Football team	1.75	0.15

Average does not tell you the value of each piece of data.
The shortest person could be in the basketball team – you cannot tell.

Let's get started

In this section you will:
- choose the correct values to find the range
- find the mean and range from a set of data
- interpret and communicate results from calculations using the mean or range

Take a look: Average running times

Ben and Alexi are talking about who is the faster runner.
The table gives the data from the last seven races.

Q Use the data to decide who you think is the faster runner.

| | 100 m times (in seconds) | | | | | | |
Race	**1**	**2**	**3**	**4**	**5**	**6**	**7**
Ben	12.8	13.2	12.9	12.6	12.9	12.8	13.0
Alexi	12.7	13.0	13.0	12.9	12.9	12.9	12.9

Here's a possible solution:

Mean time for Ben = 12.89 seconds

Mean time for Alexi = 12.90 seconds

Range for Ben = 0.6

Range for Alexi = 0.3

The mean suggests that there is no real difference between Ben and Alexi's times. However, Alexi is more consistent as his range is smaller.

R They have beaten each other three times. They have run the race in the same time once. So you need to use their average scores
The most sensible average to use here is the mean
The range will show if a runner has consistent times

A You add up Ben's times and divide by 7

A Find the difference between Ben's slowest (13.2 sec) and fastest (12.6 sec) times

I Both means round to 12.9 seconds

Have a go

1   Mrs Gordon thinks the boys in her Maths class got better marks in the last test than the girls.

Q Use the table of results below to decide if you agree with her.

Boys' scores	49%, 67%, 90%, 56%, 75%, 65%, 78%, 56%, 89%, 31%, 72%, 61%
Girls' scores	56%, 45%, 78%, 67%, 54%, 67%, 70%, 65%, 51%, 43%

2   Tom is a fisherman. He can go fishing at one of two reservoirs, Spin Water and Lin Mere.

The table below gives details of the number of fish caught per visit at both reservoirs over the past year.

Q Advise Tom on the reservoir to choose.
Give full reasons for your choice.

Spin Water		Lin Mere	
Number of fish caught per visit	Frequency (% of visitors)	Number of fish caught per visit	Frequency (% of visitors)
0	19%	0	69%
1	45%	1	5%
2	34%	2	8%
3	2%	3	8%
4	0%	4	10%

3   Beth is captain of a women's cricket team. The team will be in the final of the County Cup. Beth wants the two opening batsmen in the final to be her best and most consistent performers.

Q Look at the batsmen's statistics for the whole year.
Decide which two batsmen Beth should choose.

Batsman	Number of innings	Mean score	Range of scores	Best score
Anna	12	31.3	10	36
Fathima	2	34.0	42	55
Donna	10	33.1	20	43
Ellie	12	49.8	40	71
Lilly	11	32.6	15	35
Val	9	29.7	4	31

We're on the way

In this section you will:
- choose appropriate measures to compare two sets of data
- compare and interpret means and ranges
- make conclusions and communicate your results

Take a look: Cinema screenings

The data below compares the mean and range of the ages of people watching two films at a cinema.

(Q) Use the information to describe the age group watching each screen.

	Screen 1	Screen 2
Mean age (years)	37.3	18.4
Range of ages	60	6

(💡) Here's a possible solution:

In Screen 2, the audience had a mean age of 18.4
They had a range in age of 6.

In Screen 1, the audience had a mean age of 37.3
They had a range of 60.
They were older on average than the people in Screen 2.
They had a much wider range of ages.

Have a go

4 The Agricultural Science Research Laboratory has been testing three types of wheat for height and yield. The data they have collected is shown in the table.

	Height of crop (cm)		Yield (g per plant)	
	Mean	Range	Mean	Range
Type A	78	6	56	3
Type B	92	21	62	5
Type C	85	4	57	7

Wheat is harvested by a machine. Each plant must grow to a similar height so that it is harvested in the most cost effective manner.

(Q) Advise a group of farmers on the best type of wheat to choose. Give full reasons for your choice.

5 Bryan pays his electricity bills to Energise every quarter (every 3 months). Energise will cut his bills by 15% if he pays his bills monthly.

Another company, TrueWarm, will charge £45 per month if Bryan pays his bills by direct debit.

The table below shows how much Bryan has paid to Energise each quarter over the past 3 years.

Q Use the information to advise Bryan on the company he should use.

Year	2007				2008				2009			
Quarter	1st	2nd	3rd	4th	1st	2nd	3rd	4th	1st	2nd	3rd	4th
Bill	£164	£145	£137	£175	£184	£117	£167	£114	£136	£151	£142	£173

6 Alison owns a small restaurant with ten tables. She has two sittings in the evening. The first sitting is at 7:30 pm. The second sitting is at 9:30 pm.
To make a profit on a Saturday evening she needs to have a mean payment of £48 per table.
The information in the table shows the payment per table one Saturday evening.

Q Use the information in the table to find out if Alison makes a profit on a Saturday evening.

	7:30 pm sitting	9:30 pm sitting
Table 1	£38	empty
Table 2	£70	£36
Table 3	empty	£45
Table 4	£26	£31
Table 5	£50	£58
Table 6	£136	empty
Table 7	£32	£98
Table 8	£65	£28
Table 9	£100	£45
Table 10	£76	£66

7 Robyn has been offered two jobs when she leaves college. Both jobs are in small companies.

The table shows the salary of each person in the two companies.

Q Use the data to find the mean and mode salary of each company.
At which company would you expect Robyn to earn the larger salary?

	Whitehead's	Atkinson's
	£100 000	£50 000
	£50 000	£40 000
	£40 000	£40 000
	£40 000	£35 000
Individual salaries	£30 000	£35 000
	£18 000	£20 000
	£18 000	£20 000
	£18 000	£20 000
	£18 000	£20 000
	£18 000	£20 000

Exam ready!

In this section you will:
- decide how to use the information given in the question
- examine relationships between sets of data
- think about the implications of your results to give advice

Take a look: Average fuel economy

Alistair wants to buy a second-hand car.
There is a petrol model and a diesel model of the car he chooses.
Alister normally drives about 12 000 miles a year. He drives on different types of roads.
The average price of petrol is £5.23 per gallon. The average price of diesel is £5.28 per gallon.

Q The table shows information on both models of the car.
Use the information about fuel costs to decide which model would be the better choice.

	Year of manufacture	Miles	Average economy in mpg	Price
1.6 Petrol Model	2008	27 680	Combined 42.2	£8695
1.6 Diesel Model	2008	28 090	Combined 62.7	£9295

Fact
When a manufacturer gives average economy in mpg (miles per gallon) they are using the mean.

Here's a possible solution:

We do not need to use the information about year of manufacture and miles driven because they are the same for both cars.

	Fuel used on 12 000 miles	Cost of fuel
1.6 Petrol Model	$\frac{12\,000}{42.2} = 284$ gallons	$284 \times 5.23 = £1485$
1.6 Diesel Model	$\frac{12\,000}{62.7} = 191$ gallons	$191 \times 5.28 = £1008$

The petrol model is £477 more expensive per year, but it is £600 cheaper to buy.

A Work out how much fuel each car would use to drive 12 000 miles

So if Alistair keeps the car for 2 years, he is better off buying the diesel model.

Have a go

8 New Town School has to choose two girls to compete in the senior district athletics championship.

The school has eight possible runners for the 100 m sprint. The school will have a practice race each week for six weeks prior to the championships.

The table shows the times of the athletes in each of the six weeks before the championship.

	Week 1	Week 2	Week 3	Week 4	Week 5	Week 6
Ashleigh	14.9 s	14.6 s	14.8 s	14.7 s	14.3 s	14.5 s
Beth	15.2 s	15.4 s	15.2 s	14.4 s	14.2 s	14.1 s
Carmel	14.3 s	14.5 s	14.2 s	14.6 s	14.7 s	14.6 s
Donna	14.1 s	14.2 s	15.7 s	did not run	14.0 s	13.9 s
Fariha	14.7 s	14.6 s	14.4 s	14.5 s	14.5 s	14.6 s
Habiba	14.2 s	14.6 s	14.7 s	14.1 s	14.7 s	14.4 s
India	13.9 s	14.4 s	14.5 s	14.3 s	14.6 s	13.8 s
Jocelyn	14.0 s	14.1 s	14.6 s	14.9 s	14.4 s	14.7 s

(Q) Use the information in the table to advise the school on the two runners to choose for the championship.

9 A company wants to test a product to see if it helps people lose weight.

The company tests the product on one group of volunteers.
Another group follows an exercise plan.
A third group uses a placebo product.

All the groups follow the same diet plan.

The table shows the results.

Fact

A placebo is something that **looks** like the real product but it isn't the product. It is given to a control group. People in the control group do **not** know they are using a placebo. The results of the control group are compared with the results of the group using the real product.

Group A		Group B		Group C	
Daily use of product		Daily exercise		Daily use of placebo product	
Age	Weight loss (kg)	Age	Weight loss (kg)	Age	Weight loss (kg)
29	1.4	53	6.0	35	0.7
45	1.6	34	5.0	42	2.5
38	−0.2	47	3.9	37	2.2
43	3.6	43	4.8	51	1.0
34	2.6	39	6.4	52	3.5
37	0	44	5.2	49	−0.7
39	0.8	46	8.1	50	5.6
46	5.9	39	2.9	36	1.9
		44	6.8		

(Q) **a)** Work out the total amount of weight lost by the people in group A.

(Q) **b)** Do the trial results suggest that the company's product helps people to lose weight?

10 Ellen has a small number of sheep.

She has some purebred Durham sheep and some
Durham/Swaledale crossbred sheep.

Every Autumn Ellen takes the lambs to market. She sells the
Durham purebred lambs for 20% more per kilogram than
the crossbred lambs.

Ellen wants to have only one breed of sheep.
The table shows some information about Ellen's sheep.

Key: D = Durham   DS = Durham/Swaledale Cross

	Type of sheep	Number of lambs	Weight of lambs when sold (kg)	
Sheep 1	D	2	25	28
Sheep 2	DS	2	19	22
Sheep 3	D	1	24	
Sheep 4	DS	2	21	26
Sheep 5	D	1	27	
Sheep 6	DS	1	26	
Sheep 7	DS	1	25	
Sheep 8	D	2	25	25
Sheep 9	DS	1	24	
Sheep 10	DS	2	18	21
Sheep 11	D	1	24	
Sheep 12	DS	2	26	25

Q **a)** Work out the average number of lambs born to each breed of sheep.

Q **b)** Use the information in the table to advise Ellen if she should have
Durham purebred sheep or Durham crossbred sheep.

Now you can:

- Find the mean, median and mode and understand that each average is useful for a different purpose
- Use the range to describe the spread within a set of data
- Use averages and range to compare two sets of data
- Solve problems and communicate results to questions where averages are an important factor

11 Probability
Know Zone

Probability is often used in everyday life. Questions such as 'What are the chances of winning the lottery?', 'What is the likelihood of rain today?', 'What are the odds on Manchester United winning the European Cup?' all involve probability. You can calculate the answers to these questions if you are given enough information.

Key facts about probability

You will need to know some key facts about probability:

- Probability is the **chance**, or **likelihood**, that an event will happen.

- The probability of an event happening is between 0 and 1.
  If an event is **certain** the probability is 1.
  If an event **cannot happen** the probability is 0.

  Example: If you throw a six-sided dice the probability of throwing 1, 2, 3, 4, 5 or 6 is 1. The probability of throwing a 7 is 0.

- You can write probabilities as fractions, decimals or percentages.

- The table shows some right and wrong ways to write probabilities.

- To calculate the probability of an **event** happening you use this equation:

Correct	Incorrect
$\frac{1}{5}$	1 out of 5
$\frac{2}{10}$	2:10
0.2	1 in 5
20%	1:5

$$\text{Probability of an event} = \frac{\text{number of ways the event can happen}}{\text{total number of possibilities}}$$

Example: A person is chosen at random. The probability that this person was born in April is $\frac{30}{365}$ because there are 365 days in a year and 30 days in April. (Assuming that it is not a leap year!)

- Probability of an event happening + probability of an event not happening = 1 (or 100%).
  Example: If the probability that it will rain tomorrow is 0.4, then the probability that it will not rain tomorrow is $1 - 0.4 = 0.6$.

- The sum of the probabilities of all possible **outcomes** equals 1 (or 100%).

In this section you will:
- learn to identify the information you need to calculate probabilities
- use a numerical scale from 0 to 1 to express and compare probabilities

Take a look: Outcome of a football match

It is estimated that there is a 60% chance that a football team will win their next match. The probability that they will lose is estimated as 30%.

(Q) Work out an estimate for the probability that the football team will draw their next match.

(💡) Here's a possible solution:

There are three possible outcomes of a football match: win, lose or draw. The sum of the three probabilities must be 100%.

60% + 30% = 90%, so the probability of a draw is 100 − 90 = 10%.

Take a look: Songs on an MP3 player

Matthew downloads 2500 songs onto his MP3 player. 75 of these songs, are by the Red Hot Chilli Peppers. Becca downloads 2000 songs onto her MP3 player and 50 of the songs are by the Red Hot Chilli Peppers.

(Q) Matthew and Becca listen to their MP3 player on shuffle. Who is more likely to hear a song by the Red Hot Chilli Peppers, Matthew or Becca?

(💡) Here's a possible solution:

The probability that Matthew will here a song by the Red Hot Chilli Peppers is $\frac{75}{2500} = 0.03$

The probability that Becca will here a song by the Red Hot Chilli Peppers is $\frac{50}{2000} = 0.025$

Matthew is **more likely** to hear a song by the Red Hot Chilli Peppers on his MP3 player because 0.03 is greater than 0.025

(R) Identify the information needed to be able to work out the probability of hearing a Red Hot Chilli Peppers song for Matthew and Becca

(I) When on shuffle all of the songs on the MP3 player are equally likely to be played

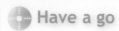 Have a go

1   A poll showed the probability that Labour will win the next general election as 34%.

(Q) What is the probability that Labour will *not* win the next general election?

2   A survey into trains being on time showed that the probability the London to Manchester train arrives on time is 0.779
    The probability that the train is early is 0.02

(Q) What is the probability that the train arrives late?

3   Out of a group of students, 32 of them travel to college by bus,
    12 of them travel by train and 16 of them walk.

(Q) A student from the group is chosen at random. What is the probability that this student travels to college by train?

4   You see two raffles advertised at a Charity Fun Day.

Raffle 1
You have a 3 in 20 chance of winning a prize!

Raffle 2
Tickets are numbered 1 to 100.
Pick a ticket ending in a 5 or a 0 and win a prize!

(Q) **In which raffle do you have a better chance of winning a prize?**

● We're on the way 1: Experimental probability

In this section you will:
◉   learn how to calculate experimental probability from the results of an experiment
◉   use probability to make conclusions

◎ Take a look: Quality control

The manufacturing industry uses **experimental probability** to assess the reliability of their products. A quality control engineer at Glow-Right Bulbs factory tested 400 light bulbs and found 6 of the light bulbs had faults.

(Q) The factory produces 50 000 light bulbs a day.
What is the probability that one light bulb has a fault?

Here's a possible solution:

Of the 400 light bulbs tested, 6 had faults,
so the probability that a light bulb has a fault is $\frac{6}{400}$
which is 0.015 (or 1.5%).

A You can write your answer as a fraction, decimal or percentage

Have a go

5 A clothing company finds that 12 in every 200 pairs of jeans do not meet their quality standards.

Q What is the probability that a pair of jeans does not meet the clothing company's quality standards?

6 Out of every 1000 cars tested in 2004, 389 cars failed the MOT test the first time they were tested.

Q What is the probability that a car fails its MOT?

7 Mike thinks a dice is biased.
He throws the dice 100 times and gets the following results.

Number	1	2	3	4	5	6
Number of times	10	15	14	19	16	26

Q Does Mike have enough evidence to decide that the dice is biased? Explain your reasoning.

⏱ We're on the way 2: Combined events

In this section you will:
- ◉ learn how to set up and use tables to list all the outcomes of two combined events
- ◉ use lists of outcomes to make conclusions

🔍 Take a look: A game using dice

A game involves throwing a pair of dice and multiplying the two numbers together to find the score.
If the number is odd you score a point. If the number is even you lose a point.
The dice are thrown 20 times.
If you have a positive number of points at the end of 20 throws you win.

Q Is the game fair?

R It helps to think of the dice as being of two different colours

💡 Here's a possible solution:

		Score on 1st dice					
		1	2	3	4	5	6
	1	1	2	3	4	5	6
	2	2	4	6	8	10	12
Score on	**3**	3	6	9	12	15	18
2nd dice	**4**	4	8	12	16	20	24
	5	5	10	15	20	25	30
	6	6	12	18	24	30	36

R First set up a table to show all the possible outcomes from throwing the two dice

The table shows that there are 6 × 6 = 36 possible outcomes.
9 of the 36 outcomes are odd. 27 of the 36 outcomes are even.
(The even outcomes are shaded in yellow).

Probability of getting an odd number $= \frac{9}{36} = 0.25$

Probability of getting an even number $= \frac{27}{36} = 0.75$

ℹ The total is either odd or even, so
Prob(odd) + Prob(even) = 1

Out of 20 throws of the dice you would expect to get:
0.25 × 20 = 5 odd numbers (scoring 5 points)
0.75 × 20 = 15 even numbers (scoring −15 points)

A Relate the outcomes to a points score

The total expected number of points at the end of
20 throws = 5 − 15 = −10.

If you score a negative number of points you lose the game, so the game is **not fair.**

ℹ Think about the expected score in relation to the fairness of the game

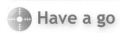

Have a go

8 A game involves tossing two coins. H = heads and T = tails.

(Q) **a)** List all of the possible outcomes when two coins are tossed.
   b) What is the probability of getting a head **and** a tail?
   c) What is the probability of both coins showing heads or both showing tails?

> **Think First!**
>
> Think about throwing two different coins, for example a 2p and a 1p coin.

9 When a football team plays a match, they can either win (W), lose (L) or draw (D).

Cheadle Town plays Stockport Rovers in a home match and an away match. Cheadle Town is equally likely to win, lose or draw each match.

(Q) What is the probability that Cheadle Town only wins one game?

10 A couple plan to have three children. Assume that the probability of a baby being born a boy or girl is the same.

(Q) **a)** Identify the possible outcomes for the gender of the children.

   b) Calculate the probability that the three children will have the same gender.

> **Think First!**
>
> Consider the order of the birth of the children.

◎ Exam ready!

In this section you will:
- ◉ use suitable forms, including tree diagrams, to represent the problem
- ◉ compare the outcomes
- ◉ show your solution to the problem clearly

◟ Take a look: Comparing chances of winning

In a school project, a student investigates the likelihood of winning in two games.

Game 1 Toss a coin three times and get three heads to win a prize.
Game 2 Three cards are labelled 1, 2 and 3. The numbers are covered, then they are shuffled. Put the cards in a line face down. Turn the cards over. If they are in numerical order you win a prize.

(Q) **a)** Identify the possible outcomes in each game.

   b) In which of the two games are you most likely to win a prize?

Here's a possible solution:

a) Game 1

First throw	Second throw	Third throw	Outcome

Game 1 tree diagram outcomes: HHH, HHT, HTH, HTT, THH, THT, TTH, TTT

> **R** The possible outcomes for this can be shown on a tree diagram

Game 2

Possible outcomes are:

1, 2, 3	1, 3, 2	2, 1, 3
2, 3, 1	3, 1, 2	3, 2, 1

> **R** **I** If the items are numbered 1, 2 and 3 you need to work out the number of ways they can be ordered

b) Game 1 The probability of winning a prize is $\frac{1}{8} = 0.125$

> **I** Out of eight possible outcomes, only one is three heads

Game 2: There are six ways to order the items, so the probability of getting them in the correct order and winning a prize is $\frac{1}{6} = 0.1666\ldots$

The highest probability of winning is $0.1666\ldots$, so you are most likely to win in Game 2.

> **R** **A** **I** Represent each of the games by making a model and then use them to work out the probabilities – don't forget to communicate your choice clearly, giving reasons

Have a go

11 A manufacturer of Best Buy sweets samples the sweets from each of three bags. The number of sweets in each bag is recorded in the table.

	Bag	Variety				Total
		Devon	Mint	Coconut	Banana	
Best Buy 400 g	A	10	13	4	15	42
	B	3	9	18	13	43
	C	6	18	5	12	41

Q **a)** Find the probability of taking a mint sweet from   **i)** bag A   **ii)** bag B   **iii)** bag C

b) Write a short statement that compares the number of sweets there might be in a bag of Best Buy sweets.

12 A medical centre makes appointments for patients in 15-minute intervals.

The centre records how many people have to wait past their appointment time.

The table shows the number of people arriving at the medical centre for appointments one week. It shows how long they had to wait past their appointment times. The times are given to the nearest minute.

	0 to 5 minutes	6 to 10 minutes	11 to 20 minutes	more than 20 minutes
Number of people	113	172	113	76

Q a) Work out the probability that a patient arriving for an appointment will have to wait more than 10 minutes.

The medical centre has a target.
Each day, the number of people who have to wait more than 20 minutes should be less than 20% of the total number of appointments.
The centre manager says that the medical centre meets this target.

Q b) Do the figures for the week of the survey support this claim?

Results Plus
Exam Tip

It is important that you set out your working clearly to show the methods you have used to arrive at your solution.

Now you can:

- Use a numerical scale to express and compare probabilities
- Identify a range of possible outcomes of combined events and record the information in tree diagrams or tables
- Make a model of a situation to work out probabilities

In the Functional Skills Mathematics exam there will be questions that need you to interpret information and make decisions, supported by reasoning. Other questions will ask you to display information using a statistical diagram or chart of your own choosing. You may also be asked to look at information using statistical and mathematical calculations.

Results Plus
Maximise your marks

Armando and Lucia own restaurants in two seaside towns on the south coast. They work out profit figures from each restaurant over the last two years. The table shows profits from April to September.

	Hastings Restaurant 35 seats		Eastbourne Restaurant 50 seats	
	2009	2010	2009	2010
April	£1500	£1300	£1700	£1700
May	£1500	£1700	£1500	£1700
June	£1700	£1900	£2000	£1900
July	£2000	£2300	£2500	£2300
August	£2400	£2400	£2400	£2300
September	£800	£800	£1600	£1300

Compare the profit for the Hastings and Eastbourne restaurants.

Exam tip

Make sure that you write your answer clearly. Give any calculations you use to inform your comments in your answer.

Student response	Exam comments
Let's look at a poor answer: Eastbourne made the most profit. Hastings made the smallest profit. I know this from the numbers in the table.	This student has made a comparison but we do not know if they have used a calculation. They may have compared any figures from the table. The response is minimal.
Let's look at an answer that will gain some marks: This year Hastings made £10 400 profit and Eastbourne made £11 200 profit. Eastbourne is better.	This student response will gain some marks. The key has been interpreted and the profit columns for 2010 have been added correctly A comparison has been made between the profits of the two catering establishments.
Let's look at a complete answer: The average monthly figures are: 		Hastings	Eastbourne
---	---	---
2009	£1650	£1950
2010	£1733	£1867 | In terms of profit per seat, in 2009: Hastings £1650 ÷ 35 = £47.14 Eastbourne £1950 ÷ 50 = £39 Eastbourne makes more money in 2009, but Hastings is more profitable in terms of the number of seats in the restaurant.	The student has made a choice of calculation and tabulated the results. They have continued by comparing the profit per seat in the restaurants in 2009. It does not matter that the student has **not** continued to analyse 2010. The data has been used to compare the figures from the two restaurants.

Exam question

The information below was collected from Secondary School pupils in the UK.
Some things the students believe about smoking are shown in the table.

Beliefs about Smoking, by age.
Percentage figures are given.

	Age				
	11 years	12 years	13 years	14 years	15 years
	Percentage who thought statement true				
Smoking helps people cope better with life	8	13	15	19	20
Smoking makes people worse at sports	82	82	83	85	85
Smoking helps people relax when nervous	49	57	69	75	79
Smokers stay slimmer than non-smokers	19	22	22	25	27
Smoking gives people self confidence	14	17	21	24	27

Do twice as many 15 year olds as 11 year olds think smoking helps people cope better with life?
Give reasons for your answer.

March 2010 Q1(b) (2 Marks)

How students answered

15% of students (0 marks)

Students who did **not** give a valid reason did **not** gain marks. These students thought about their own experiences. They did **not** look at the table and they did **not** use the data to make a conclusion.

14% of students (1 mark)

Students were asked to give a decision. Students who gave a valid reason without a decision gained only 1 mark.

71% of students (2 marks)

To gain full marks, students needed to offer a valid reason with a decision. Successful students compared 8% to 20%.

Put it into practice

1 Find situations where data is collected over a period of time and use calculations to compare the data. For example, use means, ranges, percentage changes and ratios.
   Write your answer in a way that communicates your findings clearly.

2 Imagine you have to write a short report on the data in the table above. Smoking is dangerous and harmful to your health, but more of the older students thought that smoking helps you cope better with life! How would you explain this information?

ASSESSMENT PRACTICE 3

1 Sam is writing an article on unemployment in London.

Sam collects data on the number of unemployed people in some London boroughs.

Women Unemployed	Apr 2008	Nov 2009	Men Unemployed	Apr 2008	Nov 2009
Barking and Dagenham	1086	1842	Barking and Dagenham	2637	3861
Haringey	1883	3177	Haringey	3896	5661
Harrow	752	1628	Harrow	1133	2375
Hillingdon	872	2099	Hillingdon	1879	4033
Kingston	334	902	Kingston	722	1683
Newham	1938	3180	Newham	4921	7098
Southwark	1810	3076	Southwark	4173	6160
Sutton	507	1223	Sutton	1577	3414

Q Analyse the data.
Use charts or graphs to support your analysis.

2 Doug is a security guard. He has to walk between a control room, two gates and a building.

The diagram shows the time (to the nearest minute) that it takes Doug to walk between the four parts of the site.

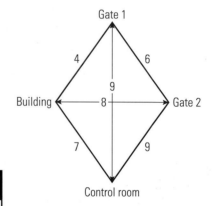

Q a) Copy and complete this table to show the time it takes Doug to walk between the different parts of the site.

	Gate 1	Building	Control room
Gate 2			
Control room			
Building			

Key: 8 = 8 m

Doug has to complete his walk between the control room, the two gates and the building every 3 hours.

Q b) Use the information in your table to advise Doug on the quickest route.

3 Amit designs a game to be played at his school fair to raise
money for charity.
He has 10 cards. Four of the cards are winning cards.
A customer will pay 20p and take a card at random.
If the card is a winning card, the customer will win 50p.
The card taken will then be put back ready for the next customer.

Amit expects 200 people to have a go at his game.
Amit works out that he is likely to make a profit.

Q Is Amit correct? You must show all your calculations.

4 The table gives information about the number of DVD players
sold by a shop each month for the last two years.

	2009	2010
January	240	195
February	183	176
March	157	143
April	148	125
May	140	108
June	132	92
July	160	132
August	117	102
September	123	90
October	136	105
November	169	129
December	252	208

The shop manager wants to compare this data.
He wants to display it in a chart or graph.

Q a) Draw a suitable chart or graph that the manager can use
to compare this data.

The shop manager needs to predict how many DVD players
he is likely to sell in 2011.
He assumes that sales will continue to fall by the same
percentage from 2010 to 2011 as they did from 2009 to 2010.

Q b) How many DVD players is the shop manager likely to sell
in 2011?

5 Penny is thinking about buying a small company. The table shows the orders for the company over 3 years.

Month	Year		
	2009	2008	2007
1	43	32	68
2	76	144	85
3	40	26	23
4	151	80	60
5	45	20	78
6	43	30	63
7	70	42	42
8	42	80	85
9	118	85	31
10	138	38	58
11	162	122	100
12	118	19	110

Key: 43 = £43 000

Q Draw a suitable chart or graph that Penny could use to compare the monthly figures for 2007, 2008 and 2009.

6 In 2007, councils in three seaside towns started trying to reduce unemployment rates in the 18–25 age group.

The graph shows the unemployment rate in the 18–25 age group for each seaside town.

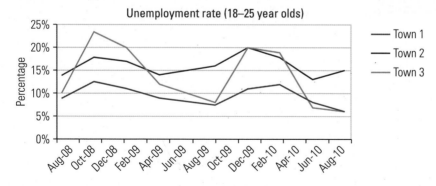

Many people visit seaside towns in the summer. There is less unemployment in the summer because more jobs are needed.

Q Use the graph to comment on how effective each council was in reducing the unemployment rates in the 18–25 age group.

7 A member of the royal family is going the visit a football club.
The club will have a welcoming ceremony for the royal visitor.
The manager of the club is going to choose a child of one of the
club's employees to give a bouquet of flowers to the royal visitor.
The manager puts the names of all the children into a box.
He will take at random one of the names. He could take the
name of:

- a child of a player
- a child of the ground staff
- a child of the admin staff

One of the football players says: 'There are three choices so the
probability that the manager will choose the child of a player is $\frac{1}{3}$'

Q Is the player right?
Explain your reasoning.

EXAM-STYLE PRACTICE

1   Yoselin wants to cover part of her kitchen wall with tiles.
The diagram shows the dimensions of the part she wants to cover.

2 m

0.75 m

(not drawn to scale)

Each tile is 10 cm square.
The tiles can be cut in half. Each half can then be used.

(Q) **a)** Show that Yoselin needs 150 tiles.

The table shows some information about the tiles Yoselin likes.

Colour of tile	Cost per tile	Cost per pack (44 tiles)
grey	£0.36	£14.88
blue	£0.33	£13.52
green	£0.31	£13.08

Yoselin wants to use grey and green tiles in the ratio 1:2
She has a budget of £55.

(Q) **b)** Can Yoselin afford to tile the wall in this way?

2   A total of five people live in Katie's house.
Katie estimates that each person uses 150 litres of water each day.
They use $\frac{1}{3}$ of this water having showers.

There are two showers in Katie's house. Katie buys new showerheads
for each shower. Each shower will now use 50% less water.

(Q) **a)** How much water will be saved each week after the new
showerheads have been fitted?

After a year, Katie thinks she has saved money by using the new showerheads.

Katie paid £24.95 for each showerhead. She paid a plumber £57 to fit the
showerheads.

Katie pays 0.28p per litre of water used.

(Q) **b)** Work out how much money Katie has saved in one year by buying
and using the showerheads.

3 Pentangle is a national car sales company. Pentangle wants to change the way that it pays its sales managers.
They will choose one of three options.

Option 1	Option 2	Option 3
Basic salary: £1000 per month	Basic salary: £1260 per month	Basic Salary: £110 per month
Sales target: 15 cars per month	Sales target: 18 cars per month	Sales target: 20 cars per month
Bonus: £100 for each extra car sold per month	Bonus: £90 for each extra car sold per month	Bonus: £180 for 1 or 2 extra cars sold £400 for 3 extra cars sold £620 for 4 extra cars sold £840 for 5 extra cars sold

'Extra cars' are cars sold above the sales target.

Sales managers at Pentangle sell an average of 24 cars per month.

(Q) Which option should Pentangle choose?

> **Think First!** You may use calculations, graphs, charts or diagrams when making your comparisons.

4 A builder is going to tile the floor of a porch. He will use a design of one black tile, two red tiles and two blue tiles, as shown in the diagram.
The builder will cover the floor completely with tiles in this design.

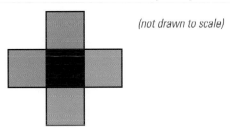

(not drawn to scale)

Each tile is a 4 cm square.
The porch is a rectangle.

1.2 m

2.4 m

(not drawn to scale)

The black tiles cost £0.25 each and the red tiles and the blue tiles cost £0.20 each.

(Q) **a)** Make a design plan of the floor.

b) Estimate the total cost of the tiles for the porch.

5 Sue and Simon start to keep a record of the alcoholic drinks they have over a period of time.

Sue's Record – Kept for 2 weeks

Friday
Went out with friends. Had two bottles of vodka mixers and four bottles of gin mixers.
Saturday
Went out. Had three bottles of pear cider.
Wednesday
Shared a bottle of white wine with a friend.
Friday
Had two bottles of pear cider.
Saturday
Had six bottles of vodka mixer.
Thursday
Shared a bottle of sparkling wine with a friend – I had most of it!

Notes
Each bottle of mixer was 1.1 units
Each bottle of pear cider was 2.8 units
Each bottle of wine was 750 ml
Each bottle of wine was 12% proof

Simon's Record – Kept for 1 week

Friday
Went down the pub and had four pints of beer.
Saturday
Had friends round to watch TV.
Drank three bottles of beer.
Sunday
Had most of a bottle of red wine.
Tuesday
Had three pints down the pub.
Wednesday
Went out with friends. Had six pints of beer.

Notes
Each pint of beer was 3 units
Each bottle of beer was 500 ml
Each bottle of beer was 5% proof
Red wine bottle was 750 ml
Red wine was 12% proof

You can use this formula to work out the number of units of alcohol in a drink.

$$\text{Units of alcohol} = \frac{V \times p}{1000}$$

where V = volume of drink (in ml)

p = percentage proof of drink

Q a) Work out how many units of alcohol there are in a 750 ml bottle of wine that is 12% proof.

Q b) How many units of alcohol do Sue and Simon drink in a typical week?

The NHS advise that:
- men should **not** have more than three to four units per day
- women should **not** have more than two to three units per day

Sue and Simon want to compare how much alcohol they drink with the NHS advice.

Q c) Use the NHS advice to comment on the number of units of alcohol Sue and Simon each drink in a week.

6 Jeeva works for a vehicle research institute. She is doing some research into how cars are driven on motorways.

You can use the formula below to work out the distance (in feet) a car needs to stop.

Stopping distance $= \dfrac{V^2}{20} + V$

Where V = speed (in miles per hour)

Q a) Work out the stopping distance for a car travelling at a speed of 60 miles per hour.

Jeeva collected data on the speed (in miles per hour) of 1000 cars. She rounded the speeds and the distances. She also collected data on the distance (in feet) of each car from the vehicle in front.

The table shows the data Jeeva collected.

Speed (miles per hour)	Mean distance from the vehicle in front (feet)	Number of cars
60	230	250
65	250	500
70	300	250

Q b) What recommendations should Jeeva make about the distance between cars travelling at the speeds shown in the table?

7 Screenwash is used to clean car windscreens. In the summer, Sam mixes one part screenwash with nine parts water.
In the winter, she mixes one part screenwash with three parts water.

Q a) Is it true that in the winter 75% of Sam's screenwash mixture is water? Explain your answer.

Sam's car has a 6-litre screenwash container. In the winter, Sam has to fill the container every 1500 miles.

Sam estimates that she will drive 6000 miles next winter.

Q b) How many 750 ml bottles of screen wash will Sam need to buy?

8 A supermarket wants to know the ages of customers using the fresh fish counter. A manager suggests two plans for recording the ages.

Plan A
Instruction: write down an estimate of the age of each customer
For example: young, or about 50, or middle aged.

Plan B
Instruction: use the following codes to record the age of each customer:
Young – Y   Middle aged – M   Old – O
For example: Y, O, M, M, O

Q a) What are the problems with each of the plans?

b) Make a plan of your own to record the age of the customers. Write down the instructions for using the plan and a method of data collection.

Level 2: Let's get started answers

Chapter 1: Number

1 £64 910, which rounded to the nearest £100 is £64 900.

2 a 10 m

  b 5 m

3 *A possible solution:*

  a C v G   C v A   C v P   A v P   A v G   P v G

  b Table of results

Name	Points
P	3
C	1
G	−1
A	−3

Chapter 2: Fractions, decimals and percentages

1 $\frac{4}{8}$ or $\frac{1}{2}$

2 25%

3 £2176 (to four significant figures)

4 Jon

5 109.545 kg (to 3 decimal places)

6 £0.0735 × 2680 + £0.0321 × (6453 − 2680) = £318.09

  Yes, the bill has been calculated correctly.

7 £688.28

8 She should sell each can for 50p because £0.40 × 1.15 = £0.46

9 Week 2 – 600 m, 12 lengths

  Week 3 – 750 m, 15 lengths (rounded up)

  Week 4 – 900 m, 18 lengths

10 £616.46

Chapter 3: Ratio and proportion

1 15 cartons of orange juice

2 2 bottles of concentrate contain 10 litres.

  Total number of parts is 8.

  This means that $\frac{3}{8}$ is concentrate and $\frac{5}{8}$ is water.

  10 litres of concentrate represents $\frac{3}{8}$

  Total number of litres of screen wash mixture possible

    = 10 ÷ 3 × 8 litres = 26.7 litres

  Capacity of mixture needed in winter = 3 × 8 litres = 24 litres

  As 26.7 is more than 24, Samil's estimate is correct.

3 Ashford group: 10, Bexhill Youth group: 6, Baslow Road group: 14

Chapter 4: Time

1 0330

2 Monday 30th April (excluding Bank Holidays)

Chapter 5: Measures

1 4.25 m

2 28 litres

3 24°C

Chapter 6: 2D representations of 3D objects

1 9 cm by 9 cm by 3.6 cm

2 27

3

4

Chapter 7: Formulae and equations

1 32°C

2 £5.22

3 £1570

Chapter 8: Area, perimeter and volume

1 12

2 5 boxes

3 £91.20

Chapter 9: Collect and represent data

1 *A possible solution:* The results are positively correlated. This shows that the more people weigh, the higher their pulse rate is.

There is one exception however. The person who weighs around 68 kg has a very low pulse rate for their weight.

2 *A possible solution:* In 1957 the family spent 52% of its income on necessities. Therefore, there was 48% left for 'luxuries'.
In 2006 the family spent 39% of its income on necessities. Therefore, there was 61% left for luxuries.
The data suggests that families in 2006 had a more luxurious lifestyle. Note: The data does not include many other factors like transport that we may regard as necessary.

3 *A possible solution:* The first graph (number of days of rain) suggests that it rains a lot more in Blaine in the summer than in Carnet. July is the month with the least amount of rainfall in Carnet. The temperature in Carnet (second graph) in July is around 26°C, which is hot but not too hot. I would therefore recommend Michael and Sarah go to Carnet in July.

4 *A possible solution:* The company profits rose from £67 000 in 2006 to £140 000 in 2008. Unfortunately, the worldwide recession meant that a much smaller profit was made in 2009.

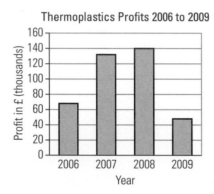

Chapter 10: Use and interpret data

1 *A possible solution:* Use the average scores to compare the results. A sensible average to use is the mean.
Mean score for the boys is 789/12 = 65.75%.
Mean score for the girls is 596/10 = 59.6%.
I agree with Mrs Gordon because the mean score for the boys is higher than the mean score for the girls.

2 *A possible solution:* At Lin Mere most people don't catch any fish at all (69%). At Spin Water the mode is 1 fish (45%). So Tom should go to Spin Water as he has more chance of catching a fish.
Another solution: If Tom is a good (or experienced) angler, he should go to Lin Mere. This is because only 2% (1 in 50) of people caught 3 or 4 fish at Spin Water. At Lin Mere, however, 18% (nearly 1 in 5) of all anglers caught 3 or more fish.

3 *A possible solution:* I would choose Ellie and Anna. Ellie has the best mean score (49.8) and the best score (71). She has also batted in 12 innings so we can be sure that the mean score is a good indication of how good she is. Anna has a similar mean score to the rest but a low range of scores over 12 innings, showing she is a consistently good performer.

Other possible choices: I would choose Donna as she has the third highest mean score and her range of 20 over 10 innings is not too high; I would choose Lilly as her mean score is similar to the rest (apart from Ellie) and her range in scores is quite low for 11 innings; I would choose Val as she is the most consistent batsmen (her range is 4) and her mean is similar to the others apart from Ellie. Note: A case can be made for all apart from Fathima, who has not played enough matches to be considered.

Chapter 11: Probability

1 66%

2 0.201

3 $\frac{12}{60}$ or $\frac{1}{5}$ or 0.2 or 20%

4 Raffle 2